Air Toxics Risk Assessment
Reference Library

Volume 2
Facility-Specific Assessment

U.S. Environmental Protection Agency
Office of Air Quality Planning and Standards
Research Triangle Park, NC

EPA-453-K-04-001B
www.epa.gov/air/oaqps

April 2004

EPA-453-K-04-001B
April 2004

**Air Toxics Risk Assessment Reference Library
Volume 2
Facility-Specific Assessment**

Prepared by:
ICF Consulting
Fairfax, Virginia

Prepared for:

Nona Smoke, Project Officer
Office of Policy Analysis and Review
Contract No. EP-D-04-005
Work Assignment No. 0-2

Rachael Schwartz, Project Officer
Clean Air Marketing Division
Contract No. 68-W-03-028
Work Assignment No. 11

Bruce Moore, Project Officer
Office of Air Quality Planning and Standards
Contract No. 68-D01-052
Work Assignment No. 0-08
Work Assignment No. 0-09

U.S. Environmental Protection Agency
Office of Air Quality Planning and Standards
Emissions Standards Division
Research Triangle Park, NC

Disclaimer

The information and procedures set forth here are intended as a technical resource to those conducting air toxics risk assessments. This facility-specific assessment document does not constitute rulemaking by the Agency, and cannot be relied on to create a substantive or procedural right enforceable by any party in litigation with the United States. As indicated by the use of non-mandatory language such as "may" and "should," it provides recommendations and does not impose any legally binding requirements.

The statutory provisions and EPA regulations described in this document contain legally binding requirements. This document is not a regulation itself, nor does not it change or substitute for those provisions and regulations. While EPA has made every effort to ensure the accuracy of the discussion in this guidance, the obligations of the regulated community are determined by statutes, regulations, or other legally binding requirements. In the event of a conflict between the discussion in this document and any statute or regulation, this document would not be controlling.

The general description provided here may not apply to a particular situation based upon the circumstances. Interested parties are free to raise questions and objections about the substance of this guidance and the appropriateness of the application of this guidance to a particular situation. EPA and other decision makers retain the discretion to adopt approaches on a case-by-case basis that differ from those described in this guidance where appropriate. EPA may take action that is at variance with the recommendations and procedures in this document and may change them at any time without public notice. This is a living document and may be revised periodically. EPA welcomes public input on this document at any time.

Reference herein to any specific commercial products, process, or service by trade name, trademark, manufacturer, or otherwise, does not necessarily constitute or imply its endorsement, recommendation, or favoring by the United States Government.

Acknowledgments

The U.S. Environmental Protection Agency's Air Toxics Risk Assessment reference library is a product of the EPA's Office of Air Quality, Planning, and Standards (OAQPS) in conjunction with EPA Regions 4 and 6 and the Office of Policy Analysis and Review. The interoffice technical working group responsible for library development includes Dr. Kenneth L. Mitchell (Region 4), Dr. Roy L. Smith (OAQPS), Dr. Deirdre Murphy (OAQPS), and Dr. Dave Guinnup (OAQPS). In addition to formal peer review, an opportunity for review and comment on Volumes 1 and 2 of the library was provided to various stakeholders, including internal EPA reviewers, state and local air agencies, and the private sector. The working group would like to thank these many internal and external stakeholders for their assistance and helpful comments on various aspects of these two books. (Volume 3 of the library is currently under development and is expected in late 2004.) The library is being prepared under contract to the U.S. EPA by ICF Consulting, Robert Hegner, Ph.D., Project Manager.

Authors, Contributors, and Reviewers

Authors

Roy L. Smith, Ph.D.
U.S. EPA OAQPS

Deirdre Murphy, Ph.D.
U.S. EPA OAQPS

Kenneth L. Mitchell, Ph.D.
U.S. EPA Region 4

External Peer Reviewers
Doug Crawford-Brown, Ph.D., University of North Carolina at Chapel Hill
Michael Dourson, Ph.D., D.A.B.T., Toxicology Excellence for Risk Assessment
Eric Hack, M.S., Toxicology Excellence for Risk Assessment
Bruce Hope, Ph.D., Oregon Department of Environmental Quality
Howard Feldman, M.S., American Petroleum Institute
Barbara Morin, Rhode Island Department of Environmental Management
Patricia Nance, M.A., M.Ed., Toxicology Excellence for Risk Assessment
Charles Pittinger, Ph.D., Toxicology Excellence for Risk Assessment, Exponent

Additional Contributors & Reviewers
John Ackermann, Ph.D., U.S. EPA Region 4
Carol Bellizzi, U.S. EPA Region 2
George Bollweg, U.S. EPA OAQPS
Pamela C. Campbell, ATSDR
Ruben Casso, U.S. EPA Region 6
Motria Caudill, U.S. EPA Region 5
Rich Cook, U.S. EPA, OTAQ
Paul Cort, U.S. EPA Region 9
David E. Cooper, Ph.D., U.S. EPA OSWER
Dave Crawford, U.S. EPA OSWER
Stan Durkee, U.S. EPA Office of Science Policy
Neal Fann, U.S. EPA OAQPS
Bob Fegley, U.S. EPA Office of Science Policy
Gina Ferreira, USEPA Region 2
Gerald Filbin, Ph.D., U.S. EPA OPEI
Danny France, U.S. EPA Region 4
Rick Gillam, U.S. EPA Region 4
Thomas Gillis, U.S. EPA OPEI
Barbara Glenn, Ph.D., U.S. EPA National Center for Environmental Research
Dave Guinnup, Ph.D., U.S. EPA OAQPS
Bob Hetes, U.S. EPA National Health and Environmental Effects Research Laboratory
James Hirtz, U.S. EPA Region 7
Ofia Hodoh, M.S., U.S. EPA Region 4

Ann Johnson, U.S. EPA OPEI
Brenda Johnson, U.S. EPA Region 4
Pauline Johnston, U.S. EPA ORIA
Stan Krivo, U.S. EPA Region 4
Deborah Luecken, U.S. EPA National Exposure Research Laboratory
Thomas McCurdy, U.S. EPA National Exposure Research Laboratory
Megan Mehaffey, Ph.D., U.S. EPA NERL
Latoya Miller, U.S. EPA Region 4
Erin Newman, U.S. EPA Region 5
David Lynch, U.S. EPA OPPTS
Ted Palma, M.S., U.S. EPA OAQPS
Michele Palmer, U.S. EPA Region 5
Solomon Pollard, Jr., Ph.D., U.S. EPA Region 4
Anne Pope, U.S. EPA OAQPS
Marybeth Smuts, Ph.D., U.S. EPA Region 1
Michel Stevens, U.S. EPA National Center for Environmental Assessment
Allan Susten, Ph.D., D.A.B.T., ATSDR
Henry Topper, Ph.D., U.S. EPA OPPTS
Pam Tsai, Sc.D., D.A.B.T., U.S. EPA Region 9
Susan R. Wyatt, U.S. EPA (retired)
Jeff Yurk, M.S., U.S. EPA Region 6

(this page intentionally left blank)

Table of Contents

List of Exhibits

(this page intentionally left blank)

Chapter I: Background

1.0   Introduction

This technical resource document describes several methods for preparing a site-specific risk assessment for a source (i.e., a single emission point within one facility), a group of sources (i.e., multiple emission points within one facility), or a group of similar facilities (e.g., within the same source category) that emit(s) toxic air pollutants. Air toxics may be emitted from power plants, factories, cars and trucks, and common household products. Sources that stay in one place are referred to as **stationary sources**. Vehicles and other moving sources of air pollutants are called **mobile sources**. *This technical resource document is intended for assessing risks associated with stationary sources of air toxics. While its primary focus is on Hazardous Air Pollutants (HAPs), this resource document can be applied to all air pollutants (with the exception of criteria air pollutants, which are assessed using different tools and methods).*

1.1   Purpose of This Document

This technical resource document is the second of a three-volume set. **Volume 1: Technical Resource Manual** discusses the overall air toxics risk assessment process and the basic technical tools needed to perform these analyses. The manual addresses both human health and ecological analyses. It also provides a basic overview of the process of managing and communicating risk assessment results. Other evaluations (such as the public health assessment process) are described to give assessors, risk managers, and other stakeholders a more holistic understanding of the many issues that may come into play when evaluating the potential impact of air toxics on human health and the environment. *Readers with a limited understanding of risk assessment are encouraged to consult Volume 1.*

Volume 2: Facility-Specific Assessment (this volume) builds on the technical tools described in Volume 1 by providing an example set of tools and procedures that can be used for source-specific or facility-specific risk assessments. Information is also provided on tiered approaches to source- or facility-specific risk analysis.

Volume 3: Community-Level Assessment builds on the information presented in Volume 1 to describe to communities how they can evaluate and reduce air toxics risks at the local level. The volume will include information on screening level and more detailed analytical approaches, how to balance the need for assessment versus the need for action, and how to identify and prioritize risk reduction options and measure success. Since community concerns and issues are often not related solely to air toxics, the document will also present readily available information on additional multimedia risk factors that may affect communities and strategies to reduce those risks. The document will provide additional, focused information on stakeholder involvement, communicating information in a community-based setting, and resources and methodologies that may play a role in the overall process. Note that EPA's Office of Pollution Prevention and Toxics has also developed a *Community Air Screening How To Manual* that will be available in 2004 and will be discussed in Volume 3 (Volume 3 will be available in late 2004).

There are multiple ways to conduct a facility/source risk assessment, and the tools and methods described in this document should not be viewed as prescriptive; nor is there a clear hierarchy of tools and methods. The specific approach selected in a risk assessment often reflects a balance between the complexity of the problem being evaluated, the uncertainty in the risk estimate that can be tolerated, and available resources. A discussion of the planned approach during planning, scoping, and problem formulation is strongly encouraged.

1.2   Intended Audience

Volume 2 is intended for three primary audiences:

•   Industries or facilities that choose to conduct site-specific risk assessments for air toxics.

•   EPA and state, local, and tribal (S/L/T) regulatory officials who may either conduct or review site-specific risk assessments as part of implementing air toxics regulatory programs.

•   Members of affected communities who wish to participate in analytical procedures related to facility-specific air toxics risk assessment.

2.0   The Facility/Source Specific Risk Assessment Process

Facility/source-specific air toxics risk assessment is the process by which the risks of adverse health or environmental impacts associated with emissions of air toxics from a defined "site" (e.g, a single facility or source) are estimated. In the context of this resource document (Volume 2) facility/source-specific risk assessment estimates are either:

•   The cumulative risk posed by the releases of HAPs from source(s) within a single source category located at a specific facility; or

> **Cumulative risk** refers to risk attributed to simultaneous exposure to multiple chemicals via a single or multiple pathways/routes

•   The cumulative risk posed by the releases of all HAPs from sources within all source categories located at a specific facility.

2.1   Facility/Source-Specific Human Health Risk Assessment

The human health risk assessment process is divided into three main phases (see Exhibit 1). These phases are described in more detail in Chapter II.

•   **Planning, scoping, and problem formulation** is performed to articulate clearly the assessment questions, state the

> Risk assessment uses scientific principles and methods to answer the following questions:
>
> •   Who is exposed to air toxics?
> •   What air toxics are they exposed to?
> •   How are they exposed?
> •   How much are they exposed to?
> •   How dangerous are specific chemicals?
> •   How likely is it that exposed people will suffer illness because of the exposures?
> •   How sure are we that our answers are correct?

quantity and quality of data needed to answer those questions, provide an in-depth discussion of how the analysis will be done, outline timing and resource considerations, identify product and documentation needs, and identify who will participate in the overall process from start to finish, along with their roles. In this example approach, planning, scoping, and problem formulation are regarded as iterative processes that allow for the assessment's plan to be adjusted as new information is obtained.

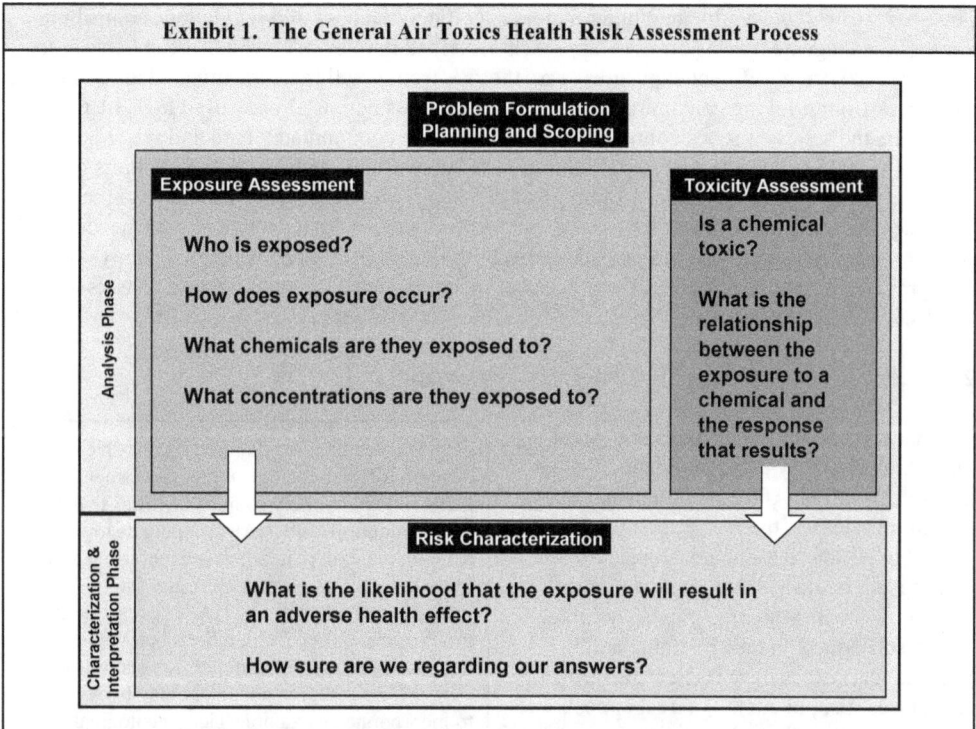

Exhibit 1. The General Air Toxics Health Risk Assessment Process

The general air toxics risk assessment process follows the risk assessment frameworks developed for EPA's *Framework for Cumulative Risk Assessment*[1] and *Residual Risk Report to Congress*[2] and is divided into three phases: (1) Planning, Scoping and Problem Formulation; (2) Analysis (consisting of exposure assessment and toxicity assessment); and (3) Risk Characterization. Volume 1 provides a more detailed discussion of these frameworks.

- **Analysis** is conducted in two steps. **Exposure Assessment** is conducted to (1) identify who is potentially exposed to air toxics; (2) identify what chemicals they may be exposed to; (3) identify how they may be exposed to those chemicals; and (4) estimate the amount of exposure. **Toxicity Assessment** considers: (1) the types of adverse health effects associated with exposure to the chemicals in question; and (2) the relationship between the amount of exposure and resulting response.

- **Risk characterization** summarizes and combines outputs of the exposure and toxicity assessments to characterize risk, both in quantitative (numerical) expressions and qualitative (descriptive) statements. Specifically, chemical-specific dose-response toxicity information is mathematically combined with modeled or monitored exposure estimates and other information about how exposure occurs to give numbers that represent the potential for the exposure to cause an adverse health outcome.

Information about risk is extremely helpful to decision makers as they try to balance the competing needs of protecting public health, sustaining economic development, evaluating issues of fairness and equity, and other factors specific to the laws and regulations controlling each risk management decision. The approach described in this document also can be used to compare risks from different exposures (e.g., the risk from breathing contaminated air compared to the risk from drinking contaminated water). Risk assessment is already used by EPA to determine the safety of foods containing pesticides, clean contaminated land and water bodies, regulate what chemicals can be imported into the United States, and set allowable limits on chemicals in our drinking water. Volume 1 of this series, the Technical Resource Manual, discusses the overall air toxics risk assessment process and the basic technical tools needed to perform these analyses. The Manual addresses both human health and ecological analyses. It also provides a basic overview of the process of managing and communicating risk assessment results.

2.2   Facility/Source-Specific Ecological Assessment

When a specific set of air toxics that persist and which also may bioaccumulate or biomagnify (PB-HAP compounds) are present in releases, the risk assessment generally will need to include consideration of exposure pathways that involve deposition of air toxics onto soil, onto plants, and into water; subsequent uptake by biota; and potential exposures to **ecological receptors** (e.g., birds, fish, plants) via direct exposure to contaminated media and/or indirect exposure through aquatic and terrestrial food chains. Air toxics ecological risk assessment is the process by which the risks of adverse impacts to ecological receptors associated with exposures to air toxics are estimated.

> EPA is investigating whether it is possible to develop a Tier 1 ecological risk assessment methodology to allow sources/facilities that emit small amounts of PB-HAP compounds to demonstrate that risk targets are met using simple look-up tables or graphs. EPA also is investigating whether any HAPs pose ecological problems in air at concentrations below the human-health based Reference Concentration (RfC). This resource document may be revised to incorporate an example Tier 1 ecological risk assessment methodology and/or an example approach for performing ecological assessments for direct exposure to air toxics in ambient air.

Congress has recognized the importance of protecting ecological receptors from adverse effects resulting from exposure to air toxics. For example, Sections 112(f)(2) through (6) of the CAA require EPA to promulgate standards beyond MACT when necessary to provide "an ample margin of safety to protect public health" and to "prevent, considering costs, energy, safety, and other relevant factors, an adverse environmental effect" (note that the requirements of other authorities may vary). The major philosophical difference between ecological and health risk management decision making is that health-based decisions intend to protect groups of individuals (e.g., population subgroups), whereas ecologically-based decisions intend to protect species and ecosystems.

The ecological risk assessment process also has three main phases that broadly correspond to the three basic phases of the human health risk assessment methodology.

- **Problem formulation**, which corresponds to the planning, scoping, and problem formulation phase of the human health risk assessment methodology;

- **Analysis**, which corresponds to the analysis phase of the human health risk assessment methodology and includes the exposure assessment and ecological effects assessment steps; and

- **Risk characterization**, which corresponds to the risk characterization phase of the human health risk assessment methodology.

> Ecological risk assessment uses scientific principles and methods to answer the following questions, which are analogous to those for human health risk assessment:
>
> - What receptors are exposed to air toxics?
> - What air toxics are they are exposed to?
> - How are they exposed (including directly and via food chains)?
> - How much are they exposed to?
> - How dangerous are the specific chemicals?
> - How likely is it that exposed receptors will suffer adverse impacts because of the exposures?
> - How sure are we that our answers are correct?

2.3 Use of Site-Specific Risk Assessments in EPA's Air Toxics Program

The results of any site-specific risk assessment can be used to support a number of activities, including:

- Development and implementation of source-specific standards and sector-based standards;

- Development and implementation of area-wide risk assessments and strategies developed by state or local air pollution control agencies to address air toxics in urban areas;

- Implementation of national air toxics assessments (NATA) activities; and

- Development of education and outreach tools.

3.0 The Layout of This Resource Document

The remainder of Volume 2 is divided into three chapters:

Chapter II: Overview and Getting Started provides an overview of risk assessment and describes several important initial steps.

- Section 1 provides an introduction to Chapter II.

- Section 2 presents an overview of the risk assessment methodology, including the basic phases and steps (planning, scoping, and problem formulation; exposure assessment; toxicity assessment; and risk characterization).

- Section 3 introduces the concept of a tiered approach, starting with a relatively simple, health-protective screening analysis and continuing with more complex and realistic analyses as needed to answer specific assessment questions.

- Section 4 describes the initial planning, scoping, and problem formulation steps that generally are followed prior to conducting the risk assessment. Three key elements are highlighted: developing a conceptual model, developing an analysis plan, and determining whether multipathway analyses are needed.

- Section 5 describes a toxicity-weighted screening analysis that can be used to focus the risk analysis on a smaller subset of HAPs that contribute the most to risk.

Chapter III: Inhalation Pathway Risk Assessment describes one set of methods and approaches that could be used to conduct human health risk assessments for inhalation exposures.

- Section 1 provides an introduction to Chapter III.

- Section 2 provides an example of how to structure a tiered assessment approach (including the use of different models at different tiers) for the inhalation analysis.

- Section 3 provides an example of how to develop the emissions inventory for an inhalation analysis.

- Section 4 describes an example of how to perform an inhalation toxicity assessment for both chronic and acute exposures.

- Section 5 describes an example risk characterization for inhalation analyses.

- Section 6 describes an example structure for a Tier 1 assessment highlighting a focus on a protective estimate of the location of the maximum exposed individual.

- Section 7 describes an example structure for a Tier 2 assessment highlighting a focus on a more realistic estimate of the highest individual risk in areas that people are believed to occupy.

- Section 8 describes an example structure for a Tier 3 assessment, highlighting the use of exposure modeling to assess variability and uncertainty in the activity patterns of the exposed population.

Chapter IV: Multipathway Risk Assessment describes one set of methods and approaches for conducting human health and ecological risk assessments for exposure pathways that involve deposition of PB-HAP compounds to soils and surface waters and subsequent uptake and ingestion exposures.

- Section 1 provides an introduction to this Chapter IV.

- Section 2 provides an example of how to structure a tiered assessment approach (including the use of different models at different tiers) for the inhalation analysis.

- Section 3 provides an example of how to develop the emissions inventory for a multipathway analysis.

- Section 4 describes an example structure for a tiered multipathway human health risk assessment for chronic exposures.

- Section 5 describes an example structure for a tiered ecological risk assessment.

This document focuses on EPA risk assessment methods and resources. Note that many state and local governments have air toxics programs. For example, California has an existing health risk assessment methodology which has been scientifically peer reviewed and has gone through extensive public comment http://www.oehha.ca.gov/air/hot_spots/index.html. State and local agencies may have existing methodologies that are acceptable for conducting facility/source-specific health risk assessments.

(this page intentionally left blank)

Chapter II: Overview and Getting Started

1.0   Introduction

This chapter provides an overview of the methodology EPA uses to quantitatively predict human health risks from emissions of HAPs at a specific facility or source in a source category. It introduces the tiered risk assessment process for facility-specific assessments. It also describes important steps in getting started, including planning, scoping, problem formulation, and conducting a toxicity-weighted emissions screening process to focus the analysis on the HAPs of greatest potential human health concern. Basic risk assessment concepts, principles, and methods are described in more detail in Volume 1 of this series. The remainder of this chapter is divided into four sections:

- Overview of Risk Assessment (Section 2);
- Concept of Tiered Assessment (Section 3)
- Planning, Scoping, and Problem Formulation (Section 4); and
- Focusing on the Most Important HAPs (Section 5).

2.0   Overview of Risk Assessment and Risk Management

The site-specific risk assessment methodology follows the risk assessment frameworks established for EPA's *Framework for Cumulative Risk Assessment*[1] and *Residual Risk Report to Congress.*[2] Volume 1 provides a more detailed discussion of these frameworks and their relationship to the general risk assessment paradigm established by the National Academy of Sciences (1983) and used throughout the federal government. The methodology includes the following components (Exhibit 2):

- Planning, scoping, and problem formulation;
- Analysis (consisting of exposure assessment and toxicity assessment); and
- Risk characterization (including qualitative or quantitative uncertainty analysis).

Any risk assessment begins with **planning and scoping**. Properly planning and scoping the risk assessment at the beginning of the project is critical to the success of the overall effort. Planning and scoping focuses on a communication step among managers, assessors, and other stakeholders regarding the purpose, scope, participants, approaches, and resources available for the risk assessment.

Problem formulation generally is conducted by both risk assessors and risk managers and focuses on two key products. The **conceptual model** identifies sources of emissions, HAPs emitted and emissions rates, the location of human and ecological receptors, potential exposure pathways/routes, and any areas (land or water) that have the potential to be contaminated from deposition of air toxics emitted from the facility/source. The conceptual model may be refined as new information becomes available during the tiered assessment process. Risk assessors use the study-specific conceptual model as a guide to help determine what types, amount, and quality of data are needed for the study to answer the questions the risk assessment has set out to evaluate. The **analysis plan** describes the specific requirements and methods to be used to obtain and analyze information on the source(s), pollutants, exposure pathways, exposed population(s), and endpoints. It also may be refined during the assessment process. This phase also may include a **toxicity-emissions weighted screening analysis** to identify which HAPs to include in the Tier 1 assessment.

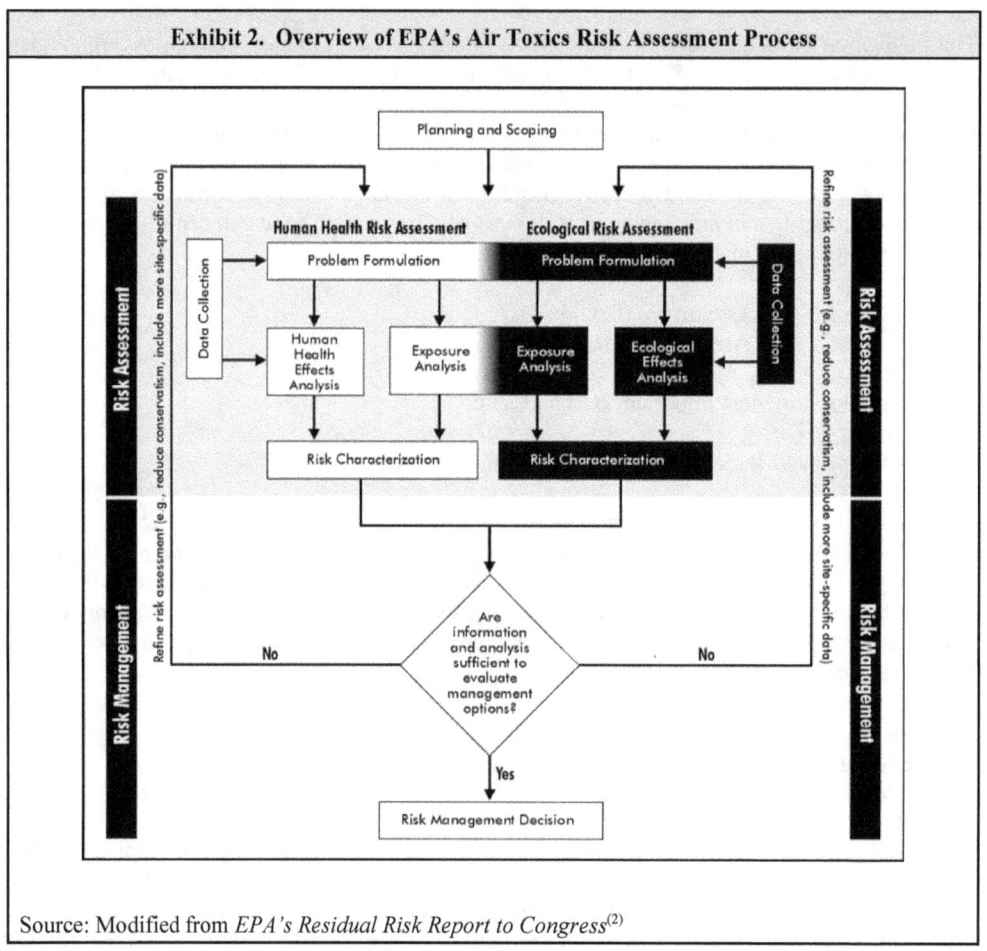

Exhibit 2. Overview of EPA's Air Toxics Risk Assessment Process

Source: Modified from *EPA's Residual Risk Report to Congress*[2]

An **exposure assessment** is conducted to identify who is potentially exposed to toxic chemicals, what chemicals they may be exposed to, and how they may be exposed to those chemicals. This often includes three substeps: (1) further characterizing the emissions sources and emissions to provide requisite model inputs; (2) performing fate and transport modeling and/or monitoring to estimate ambient concentrations of HAPs in air, water, soil, and other abiotic and biotic media (as applicable to the type of analysis); and (3) estimating inhalation exposure or oral intake (where applicable). As noted in Volume 1, estimates of exposure concentrations may be based on actual measurements (i.e. monitoring data) or and/or air quality modeling. Many studies may benefit by using some combination of modeling and monitoring, because the two approaches can be complementary.

The **toxicity assessment** component of the risk assessment process considers: (1) the types of adverse health effects associated with exposure to the chemicals in question, and (2) the relationship between the amount of exposure and resulting response.

- **Hazard identification** is the process of determining whether exposure to a chemical can cause an increase in the incidence of an adverse health effect (e.g., cancer, birth defects), and the nature and strength of the evidence for causation.

- **Dose-response assessment** is the process of quantitatively characterizing the relationship between the dose of the contaminant and the incidence of adverse health effects in the exposed population. As information on dose at the site in the body where the response occurs is rarely available, various surrogates (dose metrics) are employed, often with the assistance of biologically-based pharmacokinetic or dosimetry models to predict the dose metric from the inhalation exposure concentration or oral intake estimates. From this quantitative dose-response relationship, dose-response values are derived for use in risk characterization.[a] Most toxicity assessments are based on studies in which toxicologists expose animals to chemicals in a laboratory and extrapolate the results to humans. For some chemicals, information from actual human exposures is available (usually from studies of workplace exposures).

The **risk characterization phase** integrates the exposure and toxicity assessments to estimate risks. Cancer risk is expressed in numerical terms (e.g., 1×10^{-5} or 10 in a million) as the incremental chance an individual will develop cancer in their lifetime as a result of the exposure, and/or as the converse, the concentration corresponding to a particular level of risk. Noncancer hazard is expressed as a Hazard Quotient (HQ), the ratio of the estimated exposure to the noncancer dose-response value. For the assessment of exposures to mixtures of multiple pollutants, a Hazard Index (HI, the sum of the HQs of each chemical in a mixture) may be calculated. The individual HQs may be summed separately for chemicals that affect the same target organ/organ system or act by similar toxicological processes (e.g., the sum of the HQs for all HAPs that produce liver disease as a critical effect). Such a metric is called a Target Organ Specific Hazard Index (TOSHI). Ecological risk is often expressed as an ecological HI derived in a manner analogous to the human health HI.

Uncertainty and variability are inherent characteristics of risk assessments, and therefore the risk characterization phase includes an analysis and presentation of uncertainty and variability. Air toxics risk assessments make use of many different kinds of scientific concepts and data (e.g., exposure, toxicity, epidemiology, ecology), all of which are used to characterize the expected risk in a particular environmental context. Informed use of reliable scientific information from

---

[a]Dose-response values are numerical expressions of the relationship between a given level of exposure to an air toxic and adverse health impacts. The two most common toxicity values for inhalation exposures are the upper-bound inhalation unit risk estimates (IURs) for cancer effects and reference concentrations (RfCs) for noncancer effects (which include uncertainty factors). Chapter III, Section 4 provides a more detailed discussion of toxicity values.

many different sources is a central feature of the risk assessment process. Reliable information may or may not be available for many aspects of a risk assessment. Uncertainty and variability are inherent in the risk assessment process, and risk managers almost always must make decisions using assessments that are not as definitive in all important areas as would be desirable. Risk assessments also incorporate a variety of professional and science policy judgments (e.g., which models to use, where to locate monitors, which toxicity studies to use as the basis of developing dose-response values). Risk managers need to understand the strengths and the limitations of each assessment to communicate this information to all participants, including the public.[3] Several techniques for assessing and describing uncertainty are described more fully in Volume I (see this endnote for several key references).[4]

Risk management refers to the regulatory and other actions taken to limit or control exposures to a chemical. Risk management considers the quantitative and qualitative expressions of risk (along with attendant uncertainties) developed during the risk assessment, as well as a variety of additional information (e.g., the cost of reducing emissions or exposures, the statutory authority to take regulatory actions, and the acceptability of control options) to reach a final decision.

This document focuses on the modeling approach to estimating exposure. Ambient concentrations obtained by monitoring can be incorporated into facility/source-specific risk assessments, generally after initial (Tier 1) assessments indicate a potential for risk. As noted in Volume 1:

- The scope of monitoring is necessarily limited (i.e., spatially, temporally, and in number of pollutants measured) by resources;
- Monitoring cannot attribute exposures to sources;
- Monitoring cannot estimate exposures below the detection limit;
- Monitoring cannot predict the effects of various possible risk reduction options on future exposures; and
- Monitoring is therefore most effectively used to evaluate or further characterize modeled concentrations and exposures.

Volume 1 and EPA's NATA web site (http://www.epa.gov/ttn/atw/nata/draft6.html#secIV.B.i) provide detailed discussions of how monitoring can complement modeling in the exposure assessment.

3.0   Concept of Tiered Assessment

The human health risk assessment methodology may be based on several tiers of analysis, ranging from relatively simple, health-protective risk estimates based on limited information to complex, more realistic estimates involving more intensive data collection and calculations (see Exhibit 3).

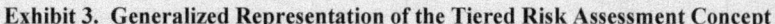

| Exhibit 3. Generalized Representation of the Tiered Risk Assessment Concept |

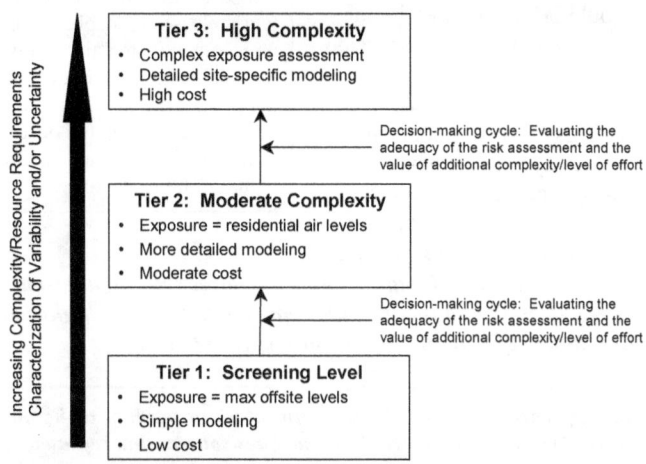

This representation, adapted from Volume III of EPA's *Risk Assessment Guidance for Superfund*,[5] depicts three tiers of analysis. Each successive tier represents more complete characterization of variability and/or uncertainty as well as a corresponding increase in complexity and resource requirements.

- Tier 1 is represented as a relatively simple, screening-level analysis using health-protective exposure assumptions [e.g., receptors are located in the area with the highest estimated concentrations] and relatively simple modeling (e.g., a model that requires few inputs, most of which can be "generic," yet health-protective).

- Tier 2 is represented as an intermediate-level analysis using more realistic exposure assumptions (e.g., use of actual receptor locations) and more detailed modeling (e.g., a model that requires additional facility/source-specific inputs).

- Tier 3 is represented as an advanced analysis, capable of using probabilistic analysis for some input variables (see Volume 1, Part VI for a discussion of these techniques) and more detailed and/or intensive modeling.

This representation does not imply that there are clear distinctions between Tiers 1, 2, and 3. For example, a series of refinements in a Tier 1 analysis might be indistinguishable from a Tier 2 analysis, or a Tier 2 analysis could incorporate probabilistic techniques.

This representation also notes the decision-making cycle that occurs between each tier. In this cycle, the existing risk assessment results are evaluated to determine whether they are sufficient for the risk management decision, and if not, what refinements to the risk assessment are needed (including moving up to the next tier).

This document presents a three-tiered approach and describe the types of models and assumptions that would be consistent with the conservativeness of these tiers. This discussion is modeled on EPA's general framework for assessing residual risk pursuant to section 112(f) of the CAA. This resource document is meant to provide an example and is not intended to prescribe a specific approach that must be used by EPA or others in a particular risk assessment activity. *In particular, various modifications to this tiered approach may be both cost-effective and appropriate, such as adding intermediate-level tiers that incorporate some features of the higher and lower tiers, or conducting iterative, more refined analyses within a given tier.*

> A **health-protective** exposure estimate is an estimate based on assumptions about exposure (e.g., release, atmospheric dispersion, contact with contaminated air) that would result in a reasonable maximum level of exposure. It is not necessarily the highest level of exposure that theoretically could occur. The term "conservative" is often used synonymously with "health protective."

> *The example Tiered approach presented in this document illustrates how an iterative risk assessment process can be used to first screen out facilities/sources that represent a relatively low risk and then focus the risk assessment on the sources and HAPs that may require emissions reduction actions. Tier 1 and Tier 2 analyses generally are an appropriate basis for deciding not to take any action or what to leave out of the next tier of analysis. Significant emissions reduction actions are likely to need a Tier 3 analysis of the specific sources and HAPs that drive the exposure and risk.*

Facility-specific assessments by regulatory agencies are often designed to answer the question: "Is the estimated risk from a facility or source category low enough to support a finding that it is of negligible regulatory concern?" If a simple, health-protective Tier 1 assessment provides a "yes" answer, there may be no need for regulatory action or further assessment. If the answer is "no," the risk manager will need to decide whether to consider a regulatory response or to refine the assessment at a higher tier. If a higher-tier assessment is performed, the decision process is repeated.

For example, a Tier 1 analysis might consist of screening-level dispersion modeling using SCREEN3. The Tier 1 analysis would typically provide a single estimate of maximum ambient air concentration that would be used to estimate inhalation risk based on an assumption that the maximum exposed individual could reside at the offsite location of maximum concentration, whether or not a person actually lived there. A Tier 2 analysis might then use the more refined Human Exposure Model, version 3 (HEM-3). HEM-3 uses the Industrial Source Complex Short-Term, version 3 (ISCST3) to provide spatially-resolved air concentrations. It combines concentration data with U.S. Census Bureau population data (2000 census) to estimate risk and hazard for the population in the most-exposed Census block. A Tier 3 analysis might couple dispersion modeling using ISCST3 with exposure modeling using $Trim.Expo_{Inhalation}$ to estimate risk and hazard for hypothetical individuals used to represent the exposed population based on combinations of demographic characteristics and activity patterns. A Tier 3 approach might also add a probabilistic exposure model to quantify the effects of variable human behavior on the exposure and risk estimates. Because the lower tier analyses are generally designed to be more health-protective, a facility-specific assessment can be performed at the lowest tier that demonstrates that applicable low-risk targets are met.

> *Determination of appropriate tiers may depend on the purpose of and regulatory context for the risk assessment. For example, if the regulatory authority for a risk management decision does not include consideration of population risk, a Tier 2 analysis might incorporate the use of a more refined air quality model to more accurately estimate maximum impact. Choice of a particular model to use within any Tier also may depend on facility/source-specific factors. For example, if downwash is particularly important at a given facility, a model other than ISCST3 may be more appropriate at Tier 2. If consideration of potential future land use is important, a model based on current land use (e.g., HEM-3) may not be the appropriate tool. Finally, the assessment may need to comply with specific S/L/T guidelines for how risk analyses should be conducted.*

4.0   Planning, Scoping, and Problem Formulation

Volume I of this reference library discusses in detail the general planning, scoping, and problem formulation process, including identifying the specific concerns; determining who will be involved in the risk assessment process (including the risk managers, risk assessment technical team, and stakeholders); communicating the purpose and scope of the risk assessment; and determining what resources are available for the risk assessment. Volume I also provides a detailed discussion of the problem formulation process, including developing a study-specific conceptual model and developing the important plans that will guide the risk assessment, including the analysis plan, and quality-related plans such as the Quality Assurance Project Plan (QAPP).

This section does not repeat the general discussion provided in Volume 1. Instead, it focuses on two aspects of planning, scoping, and problem formulation that are particularly important for facility/source-specific air toxics risk assessments (Exhibit 4): developing the conceptual model (Section 4.1); and determining whether multipathway analyses (human health ingestion and/or ecological) are applicable (Section 4.2).

Note that planning, scoping, and problem formulation activities generally continue throughout the risk assessment as new information is learned. The specific details of some activities also will vary depending on which type of analysis (i.e., human health-inhalation, human health-multipathway, ecological) and which tier of analysis is being performed.

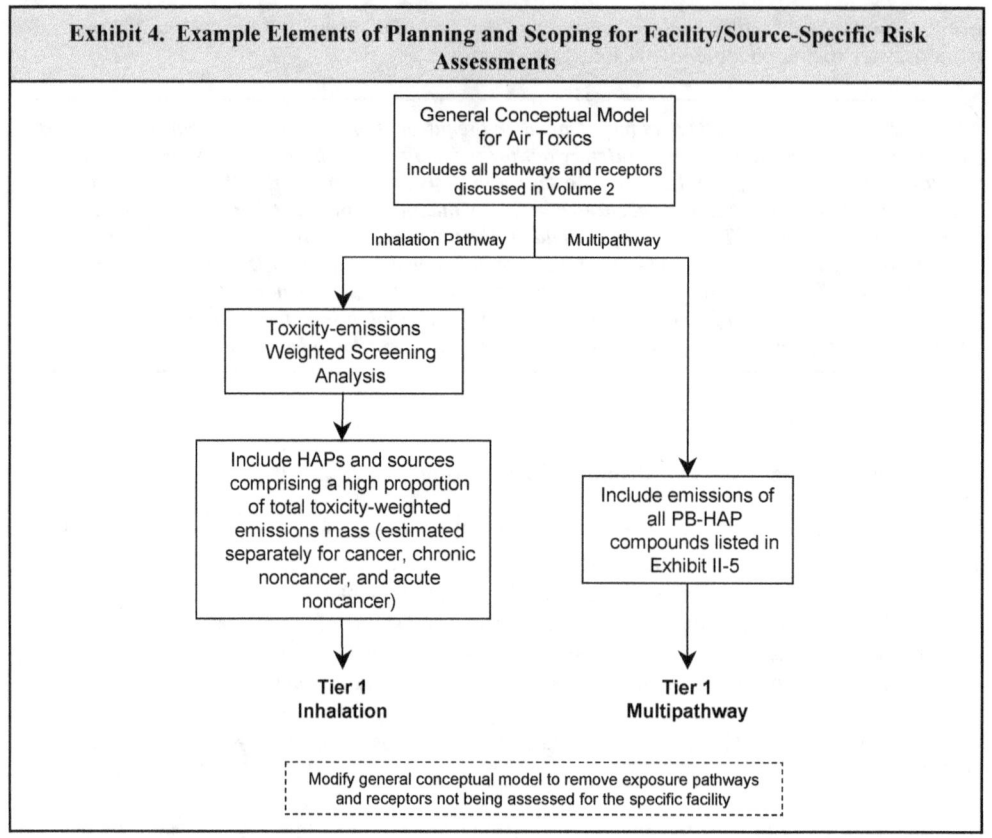

Exhibit 4. Example Elements of Planning and Scoping for Facility/Source-Specific Risk Assessments

General Conceptual Model for Air Toxics

Includes all pathways and receptors discussed in Volume 2

Inhalation Pathway

Multipathway

Toxicity-emissions Weighted Screening Analysis

Include HAPs and sources comprising a high proportion of total toxicity-weighted emissions mass (estimated separately for cancer, chronic noncancer, and acute noncancer)

Include emissions of all PB-HAP compounds listed in Exhibit II-5

Tier 1 Inhalation

Tier 1 Multipathway

Modify general conceptual model to remove exposure pathways and receptors not being assessed for the specific facility

4.1   Conceptual Model Development

The conceptual model describes the entire potential scope of the assessment, and then clarifies which pieces will be addressed. In addition to the inhalation assessments that are more or less automatic for emissions to air, the conceptual model in this example approach also considers two important chemical properties of each HAP: its persistence (P) in the environment (i.e., as determined by the HAP's half-life in air, water, soil, and sediment), and its potential to bioaccumulate (B) in plant or animal tissues (i.e., as determined by the steady-state ratio between environmental and tissue concentrations) and/or biomagnify in food chains. **PB-HAP compounds** are HAPs that are deposited onto soils and surface water, accumulate in soils, sediments, and/or biota, and normally pose a greater threat through non-inhalation pathways, especially food consumption, than by inhalation (see Section 4.2 below). Exhibit 5 provides an example of a conceptual model. The elements of a conceptual model are outlined below.

*   **Sources**. All sources (point and area) of pollutant emissions at the facility/source are identified, and a determination is made about which – if not all – of those sources need to be included in the risk assessment. In some cases, not all emission sources need to be included. For example, if a facility/source releases pollutants from a large number of sources – both

large and small – the toxicity-weighted screening analysis (TWSA) may be used to determine which sources release negligible amounts of HAPs or low toxicity pollutants that might be excluded from modeling (see Section 5 below). If data are readily available for emissions from all sources, however, the risk assessor may choose to use those data in all tiers of analyses.

- **Stressors**. All HAPs released from the sources are identified. It also is helpful at this point to characterize emissions from the sources and identify applicable dose-response values in order to perform TWSA (described in Section 5). Additional information on characterizing emissions and identifying dose-response values is provided in Section 5 below.

- **Exposure Pathways/Media/Routes**. Potential exposure pathways/routes by which the identified receptors can be exposed to the emitted HAPs are identified. The inhalation pathway commonly is included for all facility-specific assessments. Additional exposure pathways/routes, such as ingestion of animal and vegetable products raised on farms (human health), or pathways of ecological exposure, may be performed if PB-HAP compounds are present in emissions.

- **Receptors**. Human and ecological receptors that are potentially exposed to the emitted HAPs are identified, located, and characterized. This includes all areas (land or water) that have the potential to be contaminated from deposition of air toxics emitted from the facility/source. Special ecological receptors such as endangered/threatened species or wetlands are identified. The conceptual model indicates areas where the maximum exposures are expected.

> While many facility/source-specific risk assessments focus on current land use, it may be helpful to assess risks associated with potential future land use. Risk estimates based on current conditions at the facility/source and current land use could change if conditions at the facility/source change (e.g., one process is replaced by another) and/or land use changes (e.g., a housing development is built on currently undeveloped land).

- **Endpoints**. The specific human health and ecological endpoints of concern for the emitted HAPs are identified, along with specific target organs.

- **Metrics**. The specific HAP-specific and cumulative metrics used to estimate risk/hazard (e.g., cancer type, weight of evidence, target-organ-specific hazard index) are identified, along with how they will be characterized for the exposed population (e.g., central tendency, high-end, distributions).

Volume I (Chapter 6) of this reference library provides a more detailed discussion of how to perform these important steps.

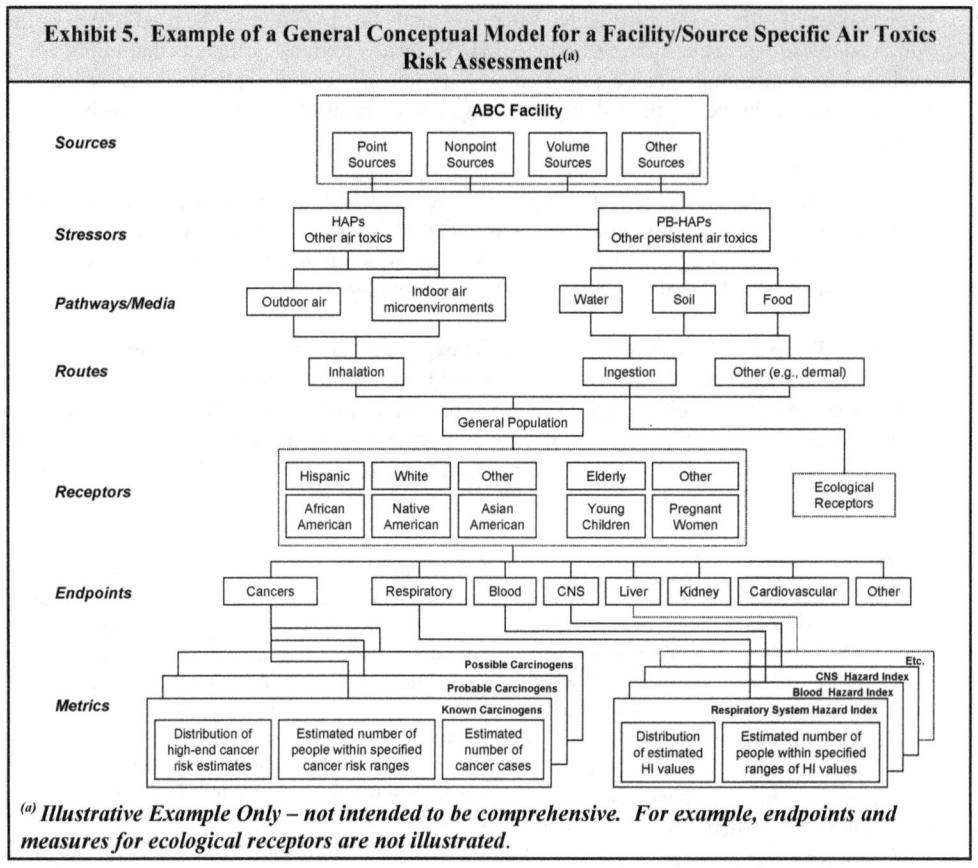

Exhibit 5. Example of a General Conceptual Model for a Facility/Source Specific Air Toxics Risk Assessment[a]

[a] *Illustrative Example Only – not intended to be comprehensive. For example, endpoints and measures for ecological receptors are not illustrated.*

In this example approach, the conceptual model is developed prior to beginning any actual estimation of exposure and risk and is iteratively modified during the planning and scoping phase, as well as during the different phases and tiers of the risk assessment, to reflect new and/or better information that is obtained. The activities that are described below are further steps in the planning and scoping phase that involve information gathering and processing. These activities will likely result in modifications to the conceptual model.

4.2   Determining Whether Multipathway Analyses are Appropriate

In this example approach, the human health inhalation pathway is evaluated in all site/source-specific risk assessments. In addition, multipathway risk assessment may be needed when specific HAPs are present in releases. **In this example approach, multipathway modeling is performed if any HAPs that persist and which also may bioconcentrate or biomagnify (i.e., PB-HAP compounds) are present in releases** (see Exhibit 6). As noted in Volume 1 (Chapter 4), all of the PB-HAP compounds are identified on one or more other Agency lists of chemicals of concern for persistence and bioaccumulation.

Exhibit 6. HAPs of Concern for Persistence and Bioaccumulation (PB-HAPs)			
PB-HAP Compound	**Pollution Prevention Priority PBTs**	**Great Waters Pollutants of Concern**	**TRI PBT Chemicals**
Cadmium compounds		X	
Chlordane	X	X	X
Chlorinated dibenzodioxins and furans	X[a]	X	X[b]
DDE	X	X	
Heptachlor			X
Hexachlorobenzene	X	X	X
Hexachlorocyclohexane (all isomers)		X	
Lead compounds	X[c]	X	X
Mercury compounds	X	X	X
Methoxychlor			X
Polychlorinated biphenyls	X	X	X
Polycyclic organic matter	X[d]	X	X[e]
Toxaphene	X	X	X
Trifluralin			X

[a] "Dioxins and furans" ("" denotes the phraseology of the source list)
[b] "Dioxin and dioxin-like compounds"
[c] Alkyl lead
[d] Benzo[a]pyrene
[e] "Polycyclic aromatic compounds" and benzo[g,h,i]perylene

It may be appropriate to consider multipathway analyses for other chemicals for which deposition may impact other media, as identified in the conceptual model for the assessment. For example, some inorganic compounds (e.g., chromium compounds, beryllium compounds) may be deposited onto plants; some semivolatile compounds may persist in soils; and in some circumstances (e.g., lack of sunlight), certain volatile organic compounds may not break down in the atmosphere. Also, some state and local agencies may have identified specific compounds or circumstances for which multipathway analyses are required. Therefore, the list of PB-HAP compounds represents a starting point for determining whether multipathway analyses are appropriate; risk assessments may need to consider additional compounds.

5.0 One Method for Focusing the Assessment on the Most Important HAPs

The initial conceptual model, which lays out the entire potential scope of the assessment, generally identifies the chemicals of potential concern (COPC) for the risk assessment. The results of the risk assessment often indicate that most of the risk is associated with a subset of the COPC. At each tier of analysis, the risk assessor might choose to reduce the number of HAPs to include in the assessment (i.e., reduce the number of COPC), with the objective (in the final tier of analysis) of identifying the subset of HAPs that drives the risk management decisions. This subset is referred to as chemicals of concern (COC). This step is optional. If the list of emitted HAPs is not long and resources are adequate, it may be appropriate to include all HAPs in the assessment.

For the Tier 1 inhalation assessment, the risk assessor might choose to reduce the number of COPC by using a simple toxicity-emissions weighted screening approach. **Note that in this example approach, all PB-HAP compounds are included in the multipathway analysis**. The **toxicity-weighted screening analysis (TWSA)** is one technique for narrowing the list of COPC for the Tier 1 inhalation risk assessment. The TWSA, a relative risk evaluation, may be calculated based on the emissions data for all HAPs released from the facility/source being assessed. A TWSA is particularly useful if there are a large number of HAPs in the facility/source emissions and there is a desire to focus the risk analysis on a smaller subset of HAPs that contribute the most to risk. A TWSA can be performed as described below.

The TWSA is intended to be entirely emissions- and toxicity-based, without considering dispersion, fate, receptor locations, and other exposure parameters. It essentially normalizes the emissions rates of each HAP to a hypothetical substance with an inhalation unit risk value of 1 per $\mu g/m^3$ (for carcinogenic effects) and/or a reference concentration (RfC) of 1 mg/m^3

> This example TWSA approach uses a cutoff of 99 percent of total toxicity-weighted emissions. *This is not intended as a suggested value*, as others (e.g., 90 or 95 percent) may be appropriate for focusing a given risk assessment on the subset of HAPs that are likely to drive the risk management decision.

(for noncancer effects). It requires emissions information as well as the applicable dose-response values (see Chapter III). This technique is especially helpful when the number of HAPs and/or the number of emission points is large. The steps for emissions-based toxicity-emissions weighted screening would include the following steps (see Exhibit 7 for an example calculation):

1. Identify all the inhalation unit risks (IURs) and RfCs for the HAPs in the facility/source emissions.
2. Determine the emission rate (e.g., tons/year) of each HAP.
3. Multiply the emission rate of each HAP by its IUR to obtain a toxicity-emissions product.
4. Rank-order the toxicity-emissions products and obtain the sum of all products.
5. Starting with the highest ranking product, proceed down the list until the cumulative sum of the products reaches a large proportion (e.g., 99 percent) of the total of the products for all the HAPs. Include in the assessment all the HAPs that contributed to this proportion of the total.

6. Repeat steps 3-5, but instead divide the emissions rate by the RfCs to obtain "noncancer equivalent tons"/year.

Exhibit 7. Example TWSA Calculation for Cancer Effects					
Air Toxic	Emissions (tons/year)	IUR	Cancer Equivalent Tons/year	Percent of Total	Cumulative Percent
1,3 butadiene	8.2×10^1	3.0×10^{-5}	2.5×10^{-3}	23.8%	23.8%
carbon tetrachloride	1.5×10^2	1.5×10^{-5}	2.2×10^{-3}	21.3%	45.1%
beryllium compounds	8.6×10^{-1}	2.4×10^{-3}	2.1×10^{-3}	19.8%	64.9%
arsenic compounds	4.2×10^{-1}	4.3×10^{-3}	1.8×10^{-3}	17.5%	82.4%
2,3,7,8-TCDD	2.0×10^{-5}	3.3×10^1	6.6×10^{-4}	6.4%	88.8%
chromium (VI) compounds	3.7×10^{-2}	1.2×10^{-2}	4.4×10^{-4}	4.3%	93.1%
polycyclic organic matter[a]	4.3	2.1×10^{-1}	3.7×10^{-4}	3.6%	96.7%
cadmium compounds	1.0×10^{-1}	1.8×10^{-3}	1.8×10^{-4}	1.8%	98.4%
formaldehyde	8.9	1.3×10^{-5}	1.2×10^{-4}	1.1%	99.5%
1,3-dichloropropene	5.2	4.0×10^{-6}	2.1×10^{-5}	0.2%	99.7%
allyl chloride	2.8	6.0×10^{-6}	1.7×10^{-5}	0.2%	99.9%
methylene chloride	1.9×10^1	4.7×10^{-7}	8.7×10^{-6}	0.1%	100.0%
benzene	9.3×10^{-2}	7.8×10^{-6}	7.3×10^{-7}	0.0%	100.0%
Total			1.0×10^{-2}	100.0%	

Heavy line denotes 99% cutoff. In this example, 1,3-dichloropropene, allyl chloride, methylene chloride, and benzene could be dropped from the cancer analysis.
[a] Cancer equivalent tons/year and IUR are based on the assumption that benzo(a)pyrene represents 5% of emissions.

Note that in subsequent tiers of analysis, a risk-based analysis can be used to further focus the assessment on the significant HAPs of concern. This approach would be similar to the TWSA, except that the risk assessor would use the Tier 1 estimates of individual cancer risk and noncancer hazard instead of toxicity-weighted emissions. The risk-based approach would include the following steps (see an example calculation in Exhibit 8):

1. Using applicable input data, run a simple dispersion and/or exposure model (with conservative assumptions) and calculate cancer risk at a selected point (e.g., maximum exposed individual location).

2. Rank-order the individual risk estimates for each emitted HAP and obtain the sum of the cancer risk.

3. Starting with the highest ranking cancer risk, proceed down the list until the individual HAPs contributing a large proportion (e.g., 99 percent) of the total risk estimate are included. Include those HAPs in subsequent tiers of analysis.

4. Repeat steps 1-3 for noncancer hazard.

Exhibit 8. Example TWSA Calculation for Noncancer Effects					
Air Toxic	Emissions (tons/year)	RfC	Noncancer Equivalent Tons/year	Percent of Total	Cumulative Percent
beryllium compounds	8.6×10^{-1}	2.0×10^{-5}	4.3×10^{4}	38.3%	38.3%
1,3 butadiene	8.2×10^{1}	2.0×10^{-3}	4.1×10^{4}	36.7%	75.0%
arsenic compounds	4.2×10^{-1}	3.0×10^{-5}	1.4×10^{4}	12.6%	87.6%
cadmium compounds	1.0×10^{-1}	2.0×10^{-5}	5.1×10^{3}	4.6%	92.1%
carbon tetrachloride	1.5×10^{2}	4.0×10^{-2}	3.7×10^{3}	3.3%	95.4%
allyl chloride	2.8	1.0×10^{-3}	2.8×10^{3}	2.5%	97.9%
formaldehyde	8.9	9.8×10^{-3}	9.1×10^{2}	0.8%	98.7%
2,3,7,8-TCDD	2.0×10^{-5}	4.0×10^{-8}	5.0×10^{2}	0.4%	99.1%
chromium (VI) compounds	3.7×10^{-2}	1.0×10^{-4}	3.7×10^{2}	0.3%	99.5%
toluene	1.3×10^{2}	4.0×10^{-1}	3.2×10^{2}	0.3%	99.8%
1,3-dichloropropene	5.2	2.0×10^{-2}	2.6×10^{2}	0.2%	100.0%
methylene chloride	1.9×10^{1}	1	1.9×10^{1}	0.0%	100.0%
benzene	9.3×10^{-2}	6.0×10^{-2}	1.6	0.0%	100.0%
		Total	1.1×10^{5}	100.0%	

Heavy line denotes 99% cutoff. In this example, chromium (VI) compounds, toluene, 1,3-dichloropropene, methylene chloride, and benzene could be dropped from the noncancer analysis.

Chapter III: Inhalation Pathway Risk Assessment

1.0   Introduction

This chapter describes an example three-tiered approach to conducting site-specific human health-inhalation risk assessments for air toxics. The tiers range from a relatively simple, health-protective screening analysis

> The three-tier approach described in this chapter is only one of a number of potential paradigms that may be appropriate and considered.

(Tier 1) to a complex probabilistic assessment (Tier 3). A risk assessor may decide to complete only the lowest-tier analysis that fits the purpose of the assessment (e.g., to determine that a facility's cumulative risk is lower than a risk manager's level of concern). Conversely, an assessor may choose not to complete a lower-tier analysis before completing a higher-tier analysis (e.g., the risk assessor could go directly to Tier 3).

The discussion in this chapter is divided into the following sections:

- Tiered Approach and Models Used (Section 2);
- Developing the Emissions Inventory for an Inhalation Analysis (Section 3);
- Toxicity Assessment (Section 4);
- Risk Characterization for Inhalation Exposures (Section 5);
- Tier 1 Inhalation Analysis (Section 6);
- Tier 2 Inhalation Analysis (Section 7); and
- Tier 3 Inhalation Analysis (Section 8).

2.0   Tiered Approach and Models Used

This example inhalation risk assessment methodology includes three tiers of analysis, each with example models for dispersion, exposure, and risk (see Exhibit 9).

- Tier 1 would use a model like SCREEN3, which is a screening-level Gaussian dispersion model that estimates an hourly maximum ambient concentration based on hourly emission rates. A relatively large degree of conservatism is incorporated in the SCREEN modeling procedure to provide reasonable assurance that maximum concentrations will not be underestimated.[6] Example default SCREEN3 inputs are identified for use in the absence of facility/source-specific data and are based on health-protective assumptions. The key output is an estimate of the highest modeled offsite concentration per model run, which is used as a surrogate for exposure. This point may be referred to as the point of maximum exposed individual (MEI). This concentration is combined with inhalation dose-response values (either for individual HAPs, or for HAPs combined by toxicity weighting) to calculate an estimate of the cumulative cancer risk and noncancer hazard to the most exposed individual. Note that SCREEN3 predicts 1-hour concentrations and requires a conversion factor in order to do the risk assessment for chronic effects (see Section 6.2.1 below).

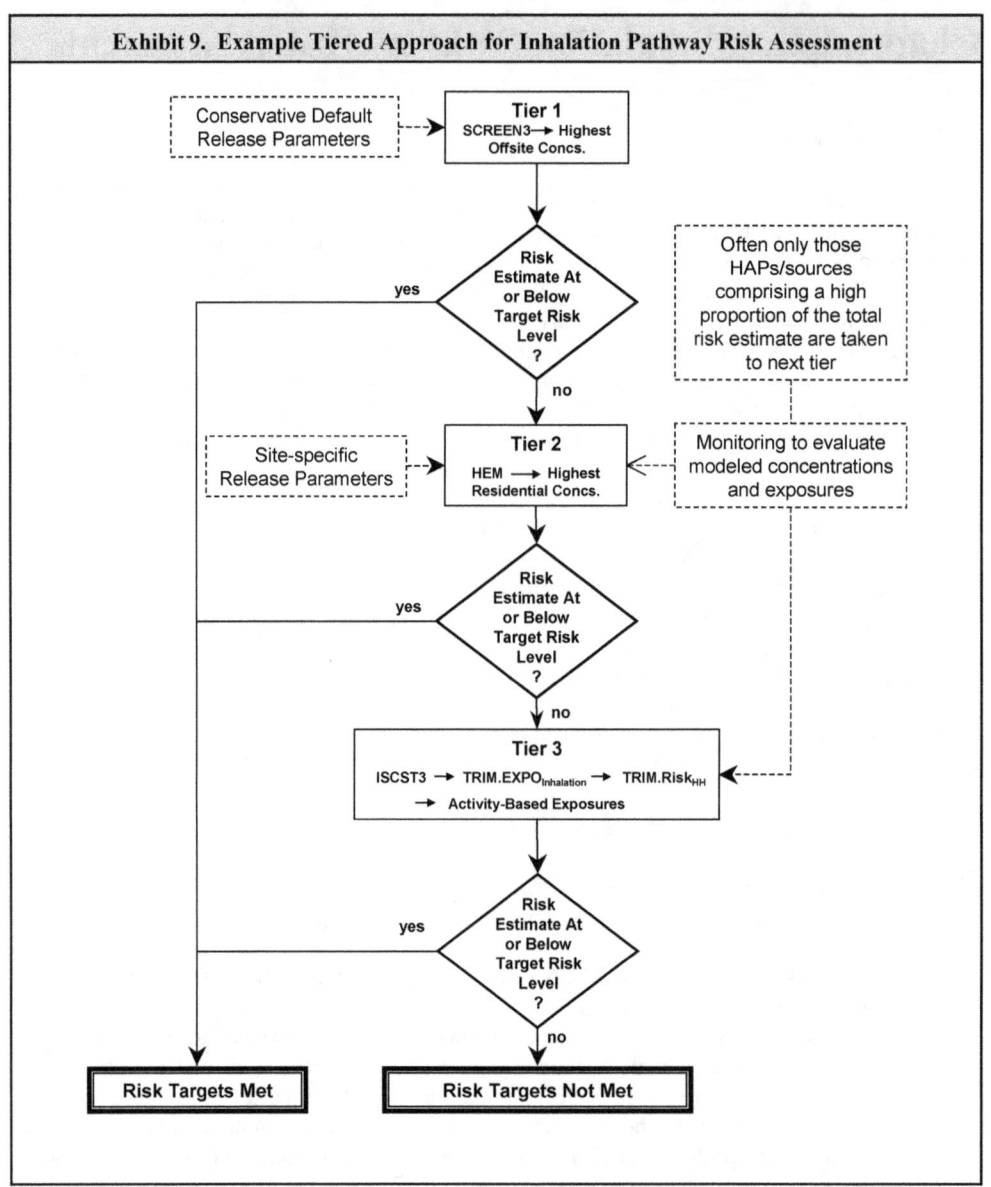

Exhibit 9. Example Tiered Approach for Inhalation Pathway Risk Assessment

Conservative Default Release Parameters

Tier 1
SCREEN3 → Highest Offsite Concs.

Risk Estimate At or Below Target Risk Level ?

Often only those HAPs/sources comprising a high proportion of the total risk estimate are taken to next tier

Site-specific Release Parameters

Tier 2
HEM → Highest Residential Concs.

Monitoring to evaluate modeled concentrations and exposures

Risk Estimate At or Below Target Risk Level ?

Tier 3
ISCST3 → TRIM.EXPO_Inhalation → TRIM.Risk_HH → Activity-Based Exposures

Risk Estimate At or Below Target Risk Level ?

yes

no

Risk Targets Met

Risk Targets Not Met

- Tier 2 would use a model like the **Human Exposure Model (HEM-3)**.[7] HEM-3 incorporates the **Industrial Source Complex-Short Term (ISCST3) model**, which is a more advanced Gaussian dispersion model that provides hourly estimates of ambient concentrations by modeling hourly emissions and meteorology. The focus in Tier 2 is on the maximum concentration to which an individual is exposed (i.e., a person who resides at the point of maximum exposure). This point may be referred to as the point of the maximum individual risk (MIR). This is done within HEM-3 using the internal point of the Census block with the highest modeled exposure or risk. It may be appropriate to consider available ambient air measurement data or conduct monitoring at Tier 2 to evaluate or further characterize modeled concentrations at specific receptor locations.

> **Key Locations for Estimating Chronic EC**
>
> - **Point of maximum modeled concentration**. The location where the maximum modeled ambient concentration occurs, regardless of whether there is a person there or not. This would provide a health-protective estimate of exposure unless someone actually lives there. This point may be referred to as the point of the **maximum exposed individual (MEI)**.
>
> - **Point of maximum modeled concentration at an actual receptor location**. The *populated* location with the highest modeled ambient concentration. This location may be an actual residence, a location within a Census block, or some other populated area. In some cases the risk assessor may decide to consider future residential use. The concentration may be interpolated between unpopulated values. This point may be referred to as the point of the maximum individual risk (MIR).

- Tier 3 would use a model like the ISCST3 model for ambient concentration estimates, but would also incorporate an exposure model like TRIM.Expo$_{Inhalation}$ to provide a more realistic estimate of individual exposures. Tier 3 would also use a model like TRIM.Risk$_{HH}$ for risk calculations. ISCST3 uses facility/source-specific source and meteorology data and provides spatially-resolved air concentrations at receptor locations. These concentrations (and/or monitoring data) are used as inputs to TRIM.Expo$_{Inhalation}$, which are used to calculate exposure to a set of hypothetical individuals, taking into account human activity patterns. TRIM.Risk$_{HH}$ combines the exposure estimates with dose-response values to calculate more refined estimates of inhalation cancer risk and noncancer hazards that account for individual differences in daily activities, including where people live and work within the assessment area. As with Tier 2, monitoring data may be helpful for evaluating the estimated ambient air concentrations (e.g., from ISCST3). Additional measurements (e.g., indoor and outdoor concentrations in specific locations where people live and work), as available, may be useful in evaluating the exposure modeling component.

As noted earlier, other models may be appropriate for these tiers (e.g., AERMOD could be used at Tier 2).

Overview of Models Cited in This Example Approach

Air Quality Models

SCREEN3. SCREEN3 is a screening-level Gaussian dispersion model that estimates an hourly maximum ambient concentration based on hourly emission rates. Short-term concentration results are used to estimate maximum annual averages via conversion factors that account for meteorological variation and the type of process that produces the emissions. Results are not direction-specific (i.e., wind direction is not taken into account). SCREEN3 calculates ambient concentrations assuming no deposition or atmospheric reactions, "worst-case" generic meteorological conditions, and flat terrain. Data requirements are relatively low. For example, SCREEN3 uses facility/source-specific data (e.g., stack height, diameter, flow rate, downwash) but does not use facility/source-specific meteorology data. Data processing requirements are low; it is easy to use for quick assessment of a single facility or source. SCREEN3 does not estimate deposition rates. SCREEN 3 is available at (http://www.epa.gov/ttn/scram/).[8]

Industrial Source Complex - Short Term (ISCST3). ISCST3 is a more advanced Gaussian dispersion model that provides hourly estimates of ambient concentrations by modeling hourly emissions and meteorology. The model includes removal effects for wet and dry deposition flux for any locations specified by the user. Data requirements are higher than for SCREEN3. For example, ISCST3 requires hourly, site-specific, processed meteorological data; and physical characteristics of emissions. Terrain information is optional. ISCST3 can accommodate variable emission rates. More expertise (e.g., specific technical and computer skills) is required to use ISCST3. Unlike SCREEN3, ISCST3 estimates annual concentrations by integrating the hourly concentrations (i.e., no multiplier is used). The ISC model is available at (http://www.epa.gov/ttn/scram/).[9]

Exposure and Risk Models

Human Exposure Model (HEM-3). HEM-3 is designed to screen major stationary sources of air pollutant emissions efficiently, ranking the sources according to the potential cancer risks and noncancer hazard associated with long-term (annual) average exposure concentrations. The current version is implemented on a Windows platform for ease of use. HEM-3 contains the Gaussian atmospheric dispersion model ISCST3 (with included meteorological data), and U.S. Census Bureau population data (2000) at the Census block level. A limited amount of source data are required as model inputs (e.g., pollutant emission rates, facility/source location, height of the emission release, stack gas exit velocity, stack diameter, temperature of the off-gases, pollutant properties and source location). HEM-3 estimates the magnitude and distribution of ambient air concentrations of pollutant in the vicinity of each source. The model usually estimates these concentrations within a radial distance of 50 kilometers (30.8 miles) from the source. Exposure concentrations for the residents of each Census block are assumed to be the outdoor concentration at the Census block "internal point." This actually represents a surrogate for exposure, as important exposure variables (e.g., indoor-outdoor concentration differences, human mobility patterns, residential occupancy period, breathing rates) are not explicitly addressed. Multiple facilities, including clusters of facilities, each having multiple emission points can be addressed by HEM-3. Variability and uncertainty in input data and parameters are not considered. The current version of HEM is available at http://www.epa.gov/ttn/fera/.[7]

Exposure and Risk Models (continued)

TRIM.Expo$_{Inhalation}$. TRIM.Expo$_{Inhalation}$ (also known as APEX3) uses a personal profile approach rather than a cohort simulation approach. That is, individuals are selected for simulation by selecting combinations of demographic characteristics and matching activity patterns, rather than directly selecting an activity pattern. If the selection probabilities for the demographic characteristics are the same as within the population to be simulated, this approach will provide a representative sample of population activity patterns without the need for post-simulation weighting of results. The current version (APEX3) includes a number of useful features, including automatic site selection from large (e.g., national) databases, a series of new output tables providing summary statistics, and a thoroughly reorganized method of describing microenvironments and their parameters. The model has the capability to estimate microenvironment concentration from the mass-balance method, but also provides the option of using the factors method. Most of the spatial and temporal constraints were removed or relaxed in APEX3. The model's spatial resolution is flexible enough to allow for the use of finely resolved modeled air quality values, as well as sparser measured values. Averaging times for exposure concentrations are equally flexible. The current version of TRIM.Expo$_{Inhalation}$ is available at http://www.epa.gov/ttn/fera/.[10]

TRIM.Risk$_{HH}$. In TRIM.Risk$_{HH}$ estimates of human exposures are characterized with regard to potential risk using the corresponding exposure- or dose-response relationships. The output from TRIM.Risk$_{HH}$ includes documentation of the input data, assumptions in the analysis, and the results of risk calculations and exposure analysis. The current version of TRIM.Risk$_{HH}$ is available at http://www.epa.gov/ttn/fera/.

Detailed Documentation

EPA's Technology Transfer Network provides detailed information regarding individual models, including software/code for each model, user's manuals, and other support documentation.

* Documentation for dispersion models may be found on EPA's Support Center for Regulatory Air Models (SCRAM) web page (http://www.epa.gov/scram001/).

* Documentation for exposure and risk models may be found on EPA's Fate, Exposure, and Risk Analysis (FERA) web page (http://www.epa.gov/ttn/fera/).

3.0   Developing the Emissions Inventory for an Inhalation Analysis

Developing the emissions inventory includes (1) quantifying emissions/release rates; (2) quantifying other emissions parameters important for the exposure assessment (e.g., temperature, stack height); (3) identifying the chemical species of the emitted HAPs (where applicable); and (4) identifying background concentrations of the HAPs being released (in some instances). Each of these is discussed in a separate subsection below.

3.1   Quantification of Emissions Rates

The risk assessor is encouraged to use the highest quality, most detailed emissions data, even in Tier 1 assessments. Depending on the objective of the risk assessment, acceptable data may include (1) actual measured emissions from a recent high-activity, high-emission year, (2) measured emissions extrapolated to a high-activity, high-emission year, (3) facility/source-specific engineering estimates of a high-activity, high emission year (with documentation); and (4) permitted emissions, (e.g., the maximum allowed under MACT, or under a permit) with documentation that the permitted limits are not exceeded.

In this example approach, Tier 1 analyses use high-end estimates of emissions in order to ensure the assessment will produce health-protective results; in subsequent tiers of analysis, more realistic emissions data (e.g., Tier 2 analyses uses average emissions for assessing cancer risks, and Tier 3 analyses may consider other factors such as spatial and temporal characteristics of releases in greater detail). The inventories commonly include the following release parameters for each HAP released from each source: volume, schedule, emission factors, and applicable time periods. Emissions commonly represent conditions typical of a high-activity, high-emission year.

In this example approach, the risk assessments consider emissions controls in use at the facility/source. The default assumption is that the facility's sources are in

> **Highest Quality vs. High-End Emissions Data**
>
> Emissions data generally are a critical starting point for air toxics risk assessments (e.g., as input data for air quality models). Use of the highest quality emissions data reduces the overall uncertainty in the resulting risk and hazard estimates.
>
> Because Tier 1 analyses are intended to be health-protective, they often are based on reasonable high-end emissions. For example, suppose five years of emissions data from a facility/source indicate that emissions vary from year-to-year. The Tier 1 analysis might be based on the single year with the highest emissions. If the resulting analysis indicates a potential for risk, subsequent Tiers of analysis could incorporate the more detailed information on emissions patterns.

compliance with the appropriate standards and permit requirements, although it may be reasonable to modify this assumption if additional emissions controls are in place. The documentation for the assessment includes the specific emissions inventories that are used.

3.2   Quantification of Other Release Parameters

Other release parameters will be needed as inputs for the various dispersion models used in the analysis. Depending on the tier of analysis, these may include parameters such as stack height, distance from the release point to the receptor, land use types, and terrain features. In general, models used in higher tiers will need greater detail and accuracy of input parameters. Where facility-specific information is not available, health-protective defaults generally are used. Some examples are provided in this reference manual (e.g., see Exhibit 12).

3.3   Dose-Response Values for "Ambiguous" Substances

This section clarifies how some specific dose-response ambiguities can be addressed in screening and tier 1 calculations (i.e., those designed to prioritize substances and sources with health-protective estimates). For the purposes of this technical resource document, an "ambiguous" substance is a HAP for which more than one dose-response assessment value could apply. Note that there may be many other uncertainties in dose-response values for a single chemical; these uncertainties are not addressed here.

Dose-response ambiguities often arise from unspeciated emissions data. That is, where: (1) substances can exist in more than one chemical form, (2) the different forms have different toxic potencies, and (3) emissions data do not identify the forms that are emitted. Examples of HAPs with different chemical forms and toxic potencies include chromium and polycyclic organic matter (POM). A second type of dose-response ambiguity can occur when there are multiple dose-response values for the same chemical form (e.g., different values for whole-life and adult exposures, for different interpretations of the dose-response data, or for food and drinking water ingestion). HAPs with this kind of ambiguity include vinyl chloride, benzene, and manganese, respectively. A single assessment (e.g., the assessment of vinyl chloride in EPA's Integrated Risk Information System [IRIS]) may provide more than one value, and therefore risk assessors have to choose which value to use.

EPA has developed a table of recommended screening-level, chronic dose-response values (available with supporting materials on-line at http://www.epa.gov/ttn/atw/toxsource/summary.html) that resolves ambiguities of the second type (dose-response assessments that provide more than one value for the same chemical form) by listing only the values appropriate for screening. Recommendations to resolve ambiguities of the first type (unspeciated emissions data) are presented below.

> This section discusses an approach for evaluating HAPs that have presented ambiguity issues in past assessments of chronic exposure, but it may not be comprehensive. For air toxics not addressed here, and for all acute exposure scenarios, a commonly used default in lieu of sufficient information supporting an alternative is always the most protective chronic dose-response value(s):
>
> - The smallest reference concentration (RfC) and reference dose (RfD);
> - The largest inhalation unit risk estimate (IUR) and cancer potency factor (CPF).
>
> This is consistent with the general philosophy behind screening-level risk assessments, in which data gaps are routinely covered by protective assumptions. This effectively minimizes the chance of a false negative (i.e., overlooking an important risk driver in the more refined assessment that follows).

The CAA lists some HAPs as groups (e.g., POM, glycol ethers, various metal compounds) that include multiple individual compounds. Toxic potency may vary widely among compounds within each of these groups. For example, many POM compounds having between 4 and 6 aromatic rings are potent mutagens and carcinogens, whereas most other POM compounds having more or fewer rings are thought to pose relatively less noncancer hazard. Hexavalent chromium (Cr^{+6}) is a carcinogen and poses risks of other effects, whereas trivalent chromium (Cr^{+3}) is not thought to be carcinogenic, is far less toxic, and is essential in the diet. However, the Toxics Release Inventory (TRI) and the National Emissions Inventory (NEI) report much of the total emitted mass of these and other ambiguous HAPs only as "total POM," "chromium compounds," or similar total values.

Exhibit 10 provides specific dose-response recommendations for screening-level risk assessments that use unspeciated HAP data. These dose-response values cover chronic exposures by ingestion (appropriate for PB-HAPs) and inhalation, and include both cancer and noncancer effects. They are based on either (1) the most toxic chemical compound or valence within the HAP group, or (2) a high-end estimate of the toxicity of mixtures emitted from different source categories. In this example approach, these recommendations are considered general-purpose screening-level defaults, to be used unless speciation information (e.g., source-specific monitoring data, EPA-approved emission factors) is available for the sources in the assessment.

3.4   Identification of Background Concentrations

For the assessment of direct source impacts, background concentrations of the released HAPs generally are not explicitly considered. However, analysis of site-specific background concentrations may be included as part of higher-tier risk assessments to place source-related risks in context with similar risks from other sources (or if total exposure is of concern). *Note that consideration of background may or may not be appropriate pursuant to the specific legal and regulatory authorities under which the risk assessment is being conducted*.

4.0   Toxicity Assessment

As noted in Volume 1 of this reference library (Chapter 12), toxicity assessment is accomplished in two steps: **hazard identification** and **dose-response assessment**. Although air toxics risk assessors need to understand the underlying scientific basis and uncertainties associated with dose-response values, they will usually rely on those values already developed and available in the literature.

Exhibit 10. Specific Dose-Response Recommendations for Unspeciated HAP Data		
HAP	**Issue**	**Default Assumption for Screening-Level Risk Assessment Where Speciated Data are Lacking[a]**
Chromium compounds	Hexavalent chromium (Cr^{+6}) is carcinogenic when inhaled, and has much higher noncarcinogenic potency than trivalent chromium (Cr^{+3})	All chromium compounds are 100% Cr^{+6} for inhalation
Cyanide (CN) compounds	RfDs vary among compounds; only hydrogen cyanide (HCN) has an RfC	All cyanide compounds are HCN for inhalation
Glycol ethers	Both RfDs and RfCs vary among compounds	All compounds are ethylene glycol methyl ether for inhalation
Hexachlorocyclohexane (HCH) (lindane and isomers)	IURs, CPFs, and RfCs vary among compounds; only lindane has an RfD	All HCH isomers are α-HCH for cancer risk (inhalation and ingestion) and lindane for noncancer hazard (inhalation and ingestion)
Mercury (Hg) compounds	RfC exists only for elemental mercury; RfDs vary between methyl mercury and mercuric chloride	All Hg is elemental for inhalation; for ingestion assume methylmercury for fish and $HgCl_2$ for other food items
Nickel (Ni) compounds	IURs exist only for nickel subsulfide and refinery dust; RfCs vary between nickel oxide and other nickel compounds	All Ni compounds are Ni_3S_2 for cancer risk (inhalation) and NiO for noncancer hazard (inhalation)
Polycyclic organic matter (POM)	IURs, CPFs, and RfDs vary widely among compounds	For cancer risk, emissions reported as total POM, total PAH, or 16-PAH are equivalent to 5% benzo[a]pyrene (BaP), and emissions reported as 7-PAH are equivalent to 18% BAP[b] (inhalation and ingestion); for noncancer hazard, all POM are pyrene

[a] Assumptions for ingestion exposure are not included for substances that are not identified as PB-HAPs
[b] High-end potency estimates for POM mixtures developed for the EPA National-Scale Air Toxics Assessment; available online as Appendix H to the Science Advisory Board review draft of the assessment at http://www.epa.gov/ttn/atw/sab/appendix-h.pdf

A major determination made during the hazard identification step concerns the potential of a chemical to cause cancer in humans. This determination, which involves considering (or weighing) all the available evidence, is called the **weight of evidence (WOE) determination**. This determination is complicated by possible inadequacies of the published studies as well as differences in body processes between people and laboratory animals. EPA's *Guidelines for Carcinogen Risk Assessment*[11] guide scientists in interpreting available studies to assess the potential human carcinogenicity of environmental pollutants. When compared with EPA's original 1986 guidelines, the 1999 interim Guidelines recommend a more comprehensive evaluation of the evidence with regard to a chemical's potential mode of action, and a more complete description of the context of a chemical's carcinogenic potential (e.g., "likely carcinogenic by inhalation and not likely carcinogenic by oral exposure"). The WOE determination now includes one of five descriptors and is accompanied by additional text that more completely summarizes EPA's interpretation of the evidence. The narrative statements consider the quality and adequacy of data and the consistency of responses induced by the agent in question (see Volume 1, Chapter 12).

Dose-response values (e.g., IURs, RfCs) are used in risk assessment to estimate the potential for adverse impacts resulting from exposure to a given concentration of a HAP. Identifying critical human health endpoints (cancer vs. non-cancer) and target organs is crucial for structuring the risk assessment, including determining what exposure pathways and routes are of potential concern, and how to sum the risks from exposure to multiple HAPs. Volume 1 of this series provides more details of this process. For each HAP included in a risk assessment, the risk assessor identifies the critical human health endpoints and target organs to ensure that cumulative risk across all HAPs is estimated in a manner consistent with risk assessment principles during the risk characterization step. The discussion of health effects criteria used by EPA in the NATA 1996 National-Scale Assessment[12] provides one example of how to identify dose-response values for a specific assessment when data from a variety of sources are available.

As noted in Volume 1 (Chapter 12), the derivation of dose-response values includes uncertainty factors and confidence levels, which vary widely among chemicals. This information should be included in the risk assessment and discussed in the risk characterization.

- **Chronic effects**. EPA/OAQPS has developed a set of recommended chronic human health dose-response values for many HAPs. This information includes the type of hazard (e.g., cancer, non-cancer) and the applicable dose-response values for each HAP (e.g., RfCs, IURs). It is presented in Appendix C of Volume 1. The most up-to-date list of default screening level dose-response values recommended by EPA for the 188 HAPs is provided at http://www.epa.gov/ttn/atw/toxsource/summary.html. This dose-response information is generally appropriate to use in any tier of a risk assessment. However, OAQPS recognizes that other independently-reviewed dose-response values may also be appropriate to use in tier 2 and 3 assessments. These values should be consistent with EPA risk assessment guidelines and agreed upon in advance with the appropriate regulatory authority for assessments that have regulatory implications. Descriptive information on the type of health hazards associated with each HAP (e.g., cancer, noncancer) may be found at http://www.epa.gov/ttn/atw/hapindex.html.

  - The **Inhalation Unit Risk (IUR)** is used to estimate inhalation cancer risk. The IUR is defined as the upper-bound excess lifetime cancer risk estimated to result from continuous exposure to an agent via inhalation per 1 $\mu g/m^3$ over a lifetime. The interpretation of unit risk would be as follows: if the IUR $= 1.5 \times 10^{-6}$ per $\mu g/m^3$, then not more than 1.5 excess tumors may be expected to develop per 1,000,000 people if exposed continuously for a lifetime to 1 μg of the chemical per cubic meter of air inhaled. The number of expected tumors may be less; it may even be none.

– The **Reference Concentration (RfC)** is used to estimate the potential for adverse effects due to chronic inhalation exposures. The RfC is defined as an estimate (with uncertainty spanning perhaps an order of magnitude) of a continuous inhalation exposure to the human population (including sensitive subpopulations) that is likely to be without an appreciable risk of deleterious effects during a lifetime. RfCs are normally expressed as milligrams per cubic meter (mg/m^3). This is generally used in EPA's health effects assessments for effects other than cancer.

• **Acute effects**. Volume 1 of this reference library (Chapter 12) identifies and describes a variety of short term, acute exposure values that may be used to assess the potential for adverse impacts due to acute inhalation exposures. As noted in Volume 1, acute criteria are developed to match specific time scales of exposure (available sources use different exposure times) and desired effect levels (e.g., no-effect or mild reversible effects). One-hour acute exposure times are commonly used for facility/source-specific risk assessments. EPA suggests comparing estimated 1-hour exposures to a range of acute dose-response values from the sources noted below. Note that these values have been developed for different purposes (e.g., some may represent mild effect levels, and some may consider economics or technical feasibility) and should be used with caution.

– **Acute Reference Exposure (ARE) values**. The ARE is an informed estimate of the highest inhalation exposure (concentration and duration) that is not likely to cause adverse effects in a human population, including sensitive subgroups, exposed to that scenario, even on an intermittent basis.[13] For these purposes, acute exposures are single continuous exposures lasting 24 hours or less; AREs may be derived for any duration of interest within that period. "Intermittent" implies sufficient time between exposures such that one exposure has no effect on the health outcome produced by the next exposure. EPA is in the process of finalizing the methodology for development of AREs.

– **1-hour Acute Exposure Guideline Levels (AEGL-1 values)**. The AEGL-1 is the airborne concentration, parts per million (ppm) or mg/m^3 of a substance above which it is predicted that the general population, including "susceptible" but excluding "hypersusceptible" individuals, could experience notable discomfort, irritation, or certain asymptomatic non-sensory effects. However, the effects are not disabling and are transient and reversible upon cessation of exposure. These values are not intended for evaluating the acute effects associated with frequent exposures; however, the use here is in comparison to the highest 1-hour exposure concentration. AEGL-1 values and supplementary information are developed by a National Advisory Committee, for use by federal, state, and local agencies and organizations in the private sector concerned with emergency planning, prevention, and response.

– **Acute Reference Exposure Levels (REL values)**. The REL is a chemical-specific acute exposure level estimate for noncancer effects (with an uncertainty spanning an order of magnitude) that is not likely to cause adverse effects in a human population after acute exposure to inhaled chemicals other than criteria air pollutants. Note that some acute RELs have different averaging periods other than one hour (e.g., 4, 5, and 6 hours). RELs are developed by the California Office of Environmental Health Hazard Assessment (OEHHA), and are available with supporting information at http://www.oehha.ca.gov/air/acute_rels/index.html.

— **Emergency Response Planning Guidelines (ERPG-1 values)**. The ERPG-1 is the maximum concentration in air below which it is believed that nearly all individuals could be exposed for up to one hour without experiencing other than mild transient adverse health effects or perceiving a clearly defined objectionable odor. These values are not intended for evaluating the acute effects associated with frequent exposures; however, the use here is in comparison to the highest 1-hour exposure concentration. ERPG-1 values are developed by the American Industrial Hygiene Association (AIHA) and are available on-line via the US Department of Energy at http://www.bnl.gov/scapa/scapawl.htm.

— **Acute Minimum Risk Levels (MRL values)**. The MRL is an estimate of human exposure to a substance that is likely to be without an appreciable risk of adverse effects (other than cancer) over a specified duration of exposure, and can be derived for acute exposures by the inhalation and oral routes. Unlike the one-hour focus of most of the other values listed here, acute MRLs are derived for exposures of 1 to 14 days duration. Acute MRLs are developed by the US Agency for Toxic Substances and Disease Registry (ATSDR), and are available at http://www.atsdr.cdc.gov/mrls.html.

— **1/10 Levels Imminently Dangerous to Life and Health (IDLH/10 Values)**. IDLH values are exactly as described, and are intended to trigger immediate evacuation of work areas. However, levels one-tenth of the IDLH tend to be generally similar to mild effect levels such as AEGL-1s or ERG-1s, and are included with EPA/OAQPS's acute values table on this basis. The IDLH/10 has been used commonly as the level of concern (superceded by ERPG and AEGL values, as available) in the Agency's emergency planning programs pursuant to the Emergency Planning and Community Right-to-Know Act (EPCRA) and Section 112(r) of the Clean Air Act. Although the use of IDLH/10 values is not ideal, in many cases these values represent the only readily-available acute dose-response value. IDLHs are developed by the National Institute for Occupational Safety and Health (NIOSH) as part of its mission to study and protect worker health, and are available at http://www.cdc.gov/niosh/idlh/idlh-1.html.

Derivation of Dose-Response Values

To understand the potential toxicity of a specific chemical, risk assessors must know both the type of effect it produces (the **hazard**) and the level of exposure required to produce that effect (the **dose-response relationship**). Both hazard and dose-response information are derived from available human epidemiologic data, experimental animal studies, and supporting information such as *in vitro* laboratory tests. This information can provide a quantitative estimate of the relationship between dose, the level of exposure, and response, the increased likelihood and/or severity of adverse effects. Dose-response assessment is the process of quantitatively evaluating toxicity information, characterizing the relationship between the dose of the contaminant administered and the incidence of adverse health effects in the exposed subjects (which may be animal or human) and then, as appropriate, extrapolating these results to human populations. Depending on the type of effect and the chemical, there are two types of dose-response values that traditionally may be derived: predictive cancer risk estimates, such as the **inhalation unit risk estimate (IUR)**, and the reference value, such as the **reference concentration (RfC)**. Both types of dose-response value may be developed for the same chemical, as appropriate.

An important aspect of dose-response relationships is whether the available evidence suggests the existence of a threshold. For many types of toxic responses, there is a **threshold dose** below which there are thought to be no adverse effects from exposure to the chemical. The human body has defenses against many toxic agents. Cells in human organs, especially in the liver and kidneys, break down many chemicals into less toxic substances that can be eliminated from the body. In this way, the human body can withstand some toxic insult (at doses below the threshold) and still remain healthy. Many HAPs are naturally occurring substances to which people routinely receive trace exposures at non-toxic levels.

Depending on whether a substance causes cancer and whether its dose-response curve is thought to have a threshold, EPA may use either of two approaches in a dose-response assessment. One approach produces a predictive estimate (e.g., inhalation cancer risk estimate), and the other produces a reference value (e.g., RfC). Historically, the use of a predictive estimate has been limited to cancer assessment. That is, dose-response assessments for cancer have been expressed as predictive cancer risk estimates based on an assumption that any amount of exposure poses some risk. Assessments of effects other than cancer usually have been expressed as reference values at or below which no harm is expected. Many substances have been assessed both ways: the first for cancer and the second for adverse effects other than cancer. While this use of predictive estimates for cancer and reference values for other effects is still the practice for the vast majority of chemicals, EPA now recognizes that there are chemicals for which the data support an alternate approach (see Volume 1, Chapter 12).

Epidemiologic and toxicologic data on air toxics typically result from exposure levels that are high relative to environmental levels, therefore **low-dose extrapolation** (prediction) is necessary to derive an applicable dose-response value. Low-dose extrapolation requires either information or assumptions about the type of dose-response curve likely under low dose situations. Confidence in the toxicity levels is indicated for noncarcinogens by applying uncertainty and modifying factors and by discussion of the confidence level. Confidence for carcinogens is indicated by EPA's weight of evidence evaluations. Volume 1, Chapter 12 provides a more detailed discussion of the derivation of dose-response values.

Derivation of Dose-Response Inhalation Values for Cancer Effects

The process for deriving a quantitative dose-response estimate for cancer (e.g., an IUR) involves the following four steps:

1. **Determination of the Point of Departure or POD.** The POD may be the traditional no observed adverse effect level (NOAEL) or lowest observed adverse effect level (LOAEL), or it may be a benchmark concentration (BMC) for tumorigenic effects.[a] The BMC is derived by the use of a mathematical curve-fitting model ("benchmark modeling") which uses the available dose-response data set for the effect to predict an exposure level associated with a particular experimental response level (e.g., observation of the effect in 10% of the experimental animals).

2. **Duration adjustment of the POD to a continuous exposure.** This step involves extrapolation of the POD from a discontinuous exposure scenario (e.g., animal studies routinely involve inhalation exposures of 6 hours per day, 5 days per week) to a POD for a continuous exposure scenario (as applicable to the RfC and IUR). This is usually accomplished by applying, as a default, a concentration-duration product (or $C \times t$ product) for both the number of hours in a daily exposure period and the number of days per week that the exposures are performed. More refined methods, such as physiologically-based pharmacokinetic (PBPK) modeling, may be used as data are available for a chemical.

3. **Extrapolation of the POD into its corresponding Human Equivalent Concentration (POD_{HEC}).** This conversion is done using dosimetric adjustment factors derived either using default methods specific to the particular chemical class of concern or more refined methods such as physiologically-based pharmacokinetic (PBPK) modeling.

4. **Extrapolation from the POD_{HEC} to lower doses.** Extrapolation from the POD_{HEC} to lower doses is usually necessary because observable cancer rates in laboratory or human occupational epidemiologic studies tend to be several orders of magnitude higher than cancer risk levels that society is willing to tolerate, and laboratory studies are conducted at exposures well above environmentally relevant concentrations. In the absence of a data set rich enough to support a biologically based model (i.e., a PBPK model), low-dose extrapolation is usually conducted using linear extrapolation or nonlinear extrapolation using a margin of exposure (MOE) analysis or derivation of an RfC.

A more detailed discussion is presented in Volume I (Chapter 12).

[a]While this is the general case for both cancer and non-cancer dose-response assessment, the LOAEL/NOAEL approach is not often used in cancer assessment.

Derivation of Dose-Response Inhalation Values for Noncancer Effects[a]

The process for deriving a quantitative dose-response estimate for noncancer effects (e.g., an RfC) involves the following five steps:

1. **Determination of the critical effect.** A critical effect is described as either the adverse effect that first appears in the dose scale as dose is increased, or as a known precursor to the first adverse effect. Underlying this designation is the assumption that if the critical effects are prevented, then all other adverse effects observed at higher exposure concentrations or doses are also prevented. Note that not all observed effects in toxicity studies are considered adverse effects. The identification of the critical effect(s) depends on a comprehensive review of the available data with careful consideration of the exposure conditions associated with each observed effect, so that comparisons of effect levels or potential reference values are made on a common basis.

2. **Determination of the Point of Departure or POD.** This step is identical to the corresponding step for cancer effects.

3. **Duration adjustment of the POD to a continuous exposure.** This step is identical to the corresponding step for cancer effects.

4. **Extrapolation of the POD into its corresponding Human Equivalent Concentration (POD_{HEC}).** This step is identical to the corresponding step for cancer effects.

5. **Application of Uncertainty Factors.** The RfC is an estimate derived from the POD_{HEC} for the critical effect by consistent application of uncertainty factors (UF). The UFs are applied to account for recognized uncertainties in the use of the available data to estimate an exposure concentration appropriate to the assumed human scenario. An uncertainty factor of 10, 3, or 1 is applied for each of the following extrapolations used to derive the RfC:
    *   Animal to human;
    *   Human to sensitive human populations;
    *   Subchronic to chronic;
    *   LOAEL to NOAEL; and
    *   Incomplete to complete database.

A more detailed discussion is presented in Volume I (Chapter 12).

[a]While the Agency has historically limited this approach to the assessment of effects other than cancer, EPA now recognizes that there are chemicals for which the data may support the use of this approach for all effects, including cancer (see Volume 1, Chapter 12)

The same chronic dose-response values (and range of acute dose-response values) can be used for all three tiers of inhalation risk assessments. If no value is available for a given HAP, that HAP is generally not quantitatively assessed for that effect (e.g., if there is no IUR, the HAP is not quantitatively assessed for cancer risk). For such HAPs, the risk assessor is encouraged to identify anything known qualitatively regarding the hazards posed by the HAP and its potential to contribute significantly to overall risk, and discuss it as part of the assessment's uncertainty analysis.

5.0   Risk Characterization for Inhalation Exposure

Risk characterization for inhalation exposures is relatively straightforward because the underlying dose-response values used are expressed in terms of exposure concentrations and take into consideration the complex physical and pharmacokinetic processes that influence how the chemical moves from the mouth/nose into the lungs, enters the blood stream, and reaches the target organ. Therefore, in the exposure assessment, dispersion modeling is used to estimate air concentrations, which are, in turn, used to estimate **exposure concentrations (ECs)** – ambient air concentrations of air toxics at the exposure points. In the risk characterization, the ECs are combined with the applicable dose-response values to generate the risk estimate. Tier 1 and Tier 2 assessments typically will assume that ambient concentrations and ECs are the same, but Tier 3 assessments will usually include separate exposure estimates developed by an exposure model.

Note that the risk and hazard estimates associated with a given EC are limited to the source(s) and exposure(s) included in the exposure assessment. Depending on decisions made in the planning, scoping, and problem formulation phase, background exposures and exposures from other sources may or may not be considered in a given risk assessment.

5.1   Cancer Risk

Risk assessors estimate excess lifetime cancer risk by combining the applicable EC and IUR for each HAP using the following equation:

$$Risk = EC_L \times IUR$$

where:

Risk	=	Individual cancer risk (expressed as an upper-bound risk of contracting cancer over a lifetime);
EC_L	=	lifetime estimate of continuous inhalation exposure to an individual HAP; and
IUR	=	the corresponding inhalation unit risk estimate for that HAP.

For inhalation cancer risk estimates, assessors assume some duration of exposure based on the characteristics of the exposed population and the purpose of the assessment. A lifetime, assumed by some air toxics risk assessments (e.g., for derivation of a MIR value), is 70 years by convention. In such assessments, while the air quality model may be run for a shorter time period (e.g., five years), and the emissions estimates also may be for a shorter time period (e.g., one year), the resulting EC is generally assumed to be representative of the entire exposure duration of interest (e.g., 70 years). If the exposure duration is less than 70 years (e.g., 30 years), then the EC would be adjusted proportionally (e.g., by multiplying the EC by 30/70). This is done because the risk of cancer is generally assumed to be related to total lifetime dose (i.e., the risk for an exposure of 100 g per year for 7 years is assumed to be equal to the risk for an exposure of 10 g per year for 70 years), and the IUR is based on an assumed lifetime exposure (see Volume 1, Chapter 21, and EPA's carcinogen risk assessment guidelines[11] for a more detailed discussion).[b]

In screening-level assessments of carcinogens for which there is an assumption of a linear dose-response, the cancer risks predicted for individual chemicals may be added to estimate cumulative cancer risk. This approach is based on an assumption that the risks associated with individual chemicals in the mixture are additive. The following equation estimates the predicted incremental individual cancer risk from multiple substances, assuming additive effects from simultaneous exposures to several carcinogens:

$$Risk_T = Risk_1 + Risk_2 + ... + Risk_i$$

where:

$Risk_T$ = total individual cancer risk (expressed as an upper-bound risk of contracting cancer over a lifetime)

$Risk_i$ = individual risk estimate for the ith HAP.

In more refined assessments, the chemicals being assessed may be evaluated to determine whether effects from multiple chemicals are synergistic (greater than additive) or antagonistic (less than additive), although sufficient data for this evaluation are usually lacking. In those cases where IURs are available for a chemical mixture of concern, risk characterization can be conducted on the mixture using the same procedures used for a single compound. *Estimating risks from chemical mixtures is a complex task that generally should be done by an experienced toxicologist.*

Estimates of cancer risk are usually expressed as a statistical probability represented in scientific notation as a negative exponent of 10. For example, an additional upper bound risk of contracting cancer of 1 chance in 10,000 (or one additional person in 10,000) is written as 1×10^{-4}. Sometimes an exponential notation is used; in this case it would be 1E-04. Because IURs are typically upper-bound estimates, actual risks may be lower than predicted (see Volume 1, Chapter 12).

In the risk characterization step, the weight-of-evidence for carcinogenicity is presented for each HAP assessed, and a qualitative discussion generally is included for those HAPs for which a quantitative assessment was not feasible (i.e., no IUR is available).

---

[b]EPA is currently reviewing methods for assessing cancer risk for less than lifetime exposures occurring in childhood. EPA's Draft Document *Supplemental Guidance for Assessing Cancer Susceptibility from Early-Life Exposure to Carcinogens* (http://www.epa.gov/sab/panels/sgacsrp.html) recommends a change to the current method for strong mutagens. This document is undergoing public and Science Advisory Board review and will be completed sometime in the future with consideration of that review. EPA's methods for air toxics assessments will be consistent with the final document.

Example Calculation to Estimate Cancer Risk (Hypothetical)

A Tier 1 modeling analysis was performed to estimate risk to the maximum exposed individual, assumed to reside at the point of maximum concentration for ABC Factory. Four HAPs were potentially of concern: benzene, dichloroethyl ether, formaldehyde, and cadmium compounds. Cancer risk estimates were obtained for each HAP by multiplying the estimated *annual average* Exposure Concentration (EC) by the Inhalation Unit Risk Estimate (IUR) for each HAP. The resulting upper bound cancer risk estimates ranged from 2×10^{-6} (benzene, formaldehyde) to 8×10^{-4} (dichloroethyl ether). The cancer risk estimates for each HAP were summed to obtain an estimate of total inhalation cancer risk (9×10^{-4}). Note that 97 percent of the estimated total risk results from dichloroethyl ether, and that more than 99 percent results from dichloroethyl ether and cadmium compounds. In this hypothetical example, the risk assessor would need to decide which HAPs to carry to higher tiers by weighing the small proportion of risk posed by benzene and formaldehyde against the fact that these risks nevertheless exceeded 1 in 1 million.

HAP	EC $\mu g/m^3$	IUR $1/(\mu g/m^3)$	Cancer Risk Estimate[a]	Percent of Total Risk
Benzene	0.3	7.8×10^{-6}	2×10^{-6}	< 1%
Dichloroethyl ether	2.5	3.3×10^{-4}	8×10^{-4}	97 %
Formaldehyde	0.2	1.3×10^{-4}	2×10^{-6}	< 1 %
Cadmium compounds	0.01	1.8×10^{-3}	1×10^{-5}	2 %
Total			9×10^{-4}	

[a] Standard rules for rounding apply which will commonly lead to an answer of one significant figure in both risk and hazard estimates. For presentation purposes, hazard quotients (and hazard indices) and cancer risk estimates are usually reported as one significant figure.

5.2   Chronic Noncancer Hazard

Risk assessors derive estimates of chronic noncancer hazard for each HAP by combining the applicable exposure concentration (EC) and reference concentration (RfC) for the HAP to obtain the chronic Hazard Quotient (HQ) for the HAP using the following equation:

$$HQ = EC_C \div RfC$$

where:

HQ   =   the chronic hazard quotient for an individual HAP;
EC_C =   estimate of continuous inhalation exposure to that HAP; and
RfC   =   the corresponding reference concentration for that HAP.

Note that, when calculating an HQ, it is very important to make sure that the EC and RfC are expressed in the same units. Modeled results (EC) are usually expressed in units of $\mu g/m^3$, *while RfCs (e.g., from IRIS) are usually expressed in units of mg/m³. Note also that 1 mg/m³ is equal to 1,000 μg/m³.*

For inhalation noncancer hazard estimates for facility/source-specific assessments, exposure estimates derived from a single year's emissions estimates are commonly used to represent a chronic exposure. Thus, if a model is run for a specified period (e.g., one year or five years), the resulting EC is assumed to be representative of a chronic exposure duration for the purposes of the chronic risk assessment.

Based on the definition of the RfC, a HQ less than or equal to one indicates that adverse noncancer effects are **not likely to occur**. With exposures increasingly greater than the RfC, (i.e., HQs increasingly greater than 1), the **potential for adverse effects increases**. However, note the following:

> **The HQs should not be interpreted probabilistically because the overall chance of adverse effects may not increase linearly as exposures exceed the RfC.**

Non-cancer health effects data are usually available only for individual HAPs within a mixture. In these cases, the individual HQs can be summed together to calculate a multi-pollutant hazard index (HI) using the following formula:

$$HI = HQ_1 + HQ_2 + ...+ HQ_i$$

where

HI = chronic hazard index; and
HQ = chronic hazard quotient for the ith HAP.

For screening-level assessments, a simple HI may first be calculated for all HAPs. This approach is based on the assumption that even when individual pollutant levels are lower than the corresponding reference levels, some pollutants may work together such that their potential for harm is additive and the combined exposure to the group of chemicals poses greater likelihood of harm. Some groups of chemicals can also behave antagonistically, such that combined exposure poses less likelihood of harm, or synergistically, such that combined exposure poses harm in greater than additive manner. Where the overall HI exceeds the criterion of interest, a more refined analysis is warranted. However, note the following:

Interpretation of differences among HQs across substances may be limited by differences among RfCs in their derivation and the fact that the slope of the dose-response curve above the RfC can vary widely depending on the substance, type of effect, and subpopulation exposed.

Because the assumption of dose additivity is most relevant to compounds that induce the same effect by similar modes of action, EPA guidance for chemical mixtures[14] suggests subgrouping pollutant-specific HQs by toxicological similarity of the pollutants for subsequent calculations, that is, to calculate a **target-organ-specific-hazard index (TOSHI)** for each subgrouping of pollutants. This allows for a more refined estimate of overall hazard.[c]

Example Calculation to Estimate Chronic Noncancer Hazard (Hypothetical)

A Tier 1 modeling analysis was performed to estimate chronic noncancer hazard to the maximum exposed individual, assumed to reside at the point of maximum concentration for ABC Factory. Four HAPs were potentially of concern: benzene, dichloroethyl ether, formaldehyde, and cadmium compounds. Noncancer hazard estimates were obtained for each HAP by dividing the estimated Exposure Concentration (EC) by the Inhalation Reference Concentration (RfC) for each HAP (note that the EC is expressed in units of mg/m^3 for this analysis). The resulting Hazard quotient (HQ) estimates ranged from 1×10^{-3} (formaldehyde) to 1 (cadmium compounds). Note that no RfC was available for dichloroethyl ether. The HQs for each HAP were summed to obtain an estimate of the Hazard Index (HI) of 1. Note that cadmium compounds account for 95 percent of the HI, suggesting that the other HAPs may not need further consideration (although this determination should be made in consideration of all relevant information, including uncertainties such as confidence in the exposure concentration and uncertainty factors used to derive each RfC).

HAP	EC mg/m^3	RfC (mg/m^3)	HQ[b]	Percent of HI
Benzene	6×10^{-4}	6×10^{-2}	1×10^{-2}	1 %
Dichloroethyl ether[a]	5×10^{-3}	---	---	---
Formaldehyde	4×10^{-4}	9.8×10^{-3}	1×10^{-3}	4 %
Cadmium compounds	2×10^{-5}	2×10^{-5}	1	95 %
Hazard Index (HI)			1	

[a] note that the absence of an RfC value means that we cannot quantitatively assess a HAP.
[b] Standard rules for rounding apply which will commonly lead to an answer of one significant figure in both risk and hazard estimates. For presentation purposes, hazard quotients (and hazard indices) and cancer risk estimates are usually reported as one significant figure.

Although the HI approach encompassing all chemicals in a mixture is commonly used for a screening-level study, it is important to note that application of the HI equation to compounds that may produce different effects, or that act by different mechanisms, could overestimate the potential for effects. Consequently, in a refined assessment, risk assessors commonly calculate a separate HI for each noncancer endpoint of concern those HAPs with toxicological similarity. Descriptive information on the type of noncancer health hazards associated with each HAP may be found at http://www.epa.gov/ttn/atw/hapindex.html. *Note that assessing hazards from chemical mixtures (e.g., through calculation of a TOSHI) is a complex task that generally should be done by an experienced toxicologist.*

---

[c]The all-HAP HI generally is only used in screening-level (Tier 1) analyses to identify situations where the risk assessor is confident that the estimate of hazard is unlikely to be greater than the risk management decision criterion (e.g., a HI greater than 1). The all-HAP HI often is not used in more refined analyses.

A qualitative discussion generally is included in the risk characterization for those HAPs for which a quantitative assessment was not feasible (i.e., no RfC is available).

5.3   Acute Noncancer Hazard

Risk assessors can derive estimates of acute noncancer hazard for each HAP by combining the applicable short-term exposure concentration (EC) and acute dose-response value (AV) for the HAP to obtain the acute Hazard Quotient (HQ) for the HAP using the following equation:

$$HQ_A = EC_{ST} \div AV$$

where:

$\quad HQ_A \quad = \quad$ the acute hazard quotient for an individual HAP;
$\quad EC_{ST} \quad = \quad$ estimate of short-term inhalation exposure to that HAP; and
$\quad AV \quad = \quad$ the corresponding acute dose-response value for that HAP.

Note that ambient air concentrations are calculated for an exposure duration compatible with the acute dose-response value used.

Acute noncancer health effects data are usually available only for individual HAPs within a mixture. In these cases, it may be possible to combine the individual acute HQs to calculate a multi-pollutant acute hazard index (HI) using the following formula:

$$HI_A = HQ_{A1} + HQ_{A2} + ...+ HQ_{Ai}$$

where

$\quad HI_A \quad = \quad$ acute hazard index; and
$\quad HQ_{Ai} \quad = \quad$ acute hazard quotient for the ith HAP.

Although this appears similar to the process for combining chronic HQs, the summing of acute HQs is complicated by several issues that do not pertain to chronic HQs. First, acute dose-response values have been developed for purposes that vary more widely than chronic values. Some sources of acute values define exposures at which adverse effects actually occur, while other sources develop only no-effect acute values. Second, some acute values are expressed as concentration-time matrices, while others are expressed as single concentrations for a set exposure duration. Third, some acute values may specifically consider multiple exposures, whereas others consider exposure as a one-time event. Fourth, some sources of acute values are intended to regulate workplace exposures, assuming a population of healthy workers (i.e., without children, seniors, or other sensitive individuals). Such occupational values may also consider cost and feasibility, factors that EPA considers the province of the risk manager rather than the risk assessor.

Given these differences among acute values with regard to their purposes, and the different types of acute exposure characterization that may be performed, the acute HI analysis is most informative when limited to acute values from the same source, the same level of effects, and the same duration. Analyses that mix sources, effects levels, and durations are likely to be misleading.

Risk assessors commonly evaluate acute noncancer hazard using a variety of different acute values from different sources, and discuss the resulting hazard estimates considering the purpose for which each of value was developed. This kind of evaluation should only be done by an experienced toxicologist. **The significance of these HQs and HIs would need to be considered in the context of the purpose of the risk assessment and the characteristics of the dose-response values, such as their purpose, averaging time, and health endpoints.** EPA is working to provide more comprehensive guidance on what benchmarks to rely upon and plans to develop a relevant acute benchmark methodology.

5.4   Assessment and Presentation of Uncertainty

The risk estimates used in air toxics risk assessments usually are not fully probabilistic estimates of risk but conditional estimates given a considerable number of assumptions about exposure and toxicity. Air toxics risk assessments make use of many different kinds of scientific concepts and data (e.g., exposure, toxicity, epidemiology), all of which are used to characterize the expected risk in a particular environmental context. Informed use of reliable scientific information from many different sources is a central feature of the risk assessment process. Reliable information may or may not be available for many aspects of a risk assessment. Scientific uncertainty is inherent in the risk assessment process, and risk managers almost always must make decisions using assessments that are not as definitive in all important areas as would be desirable. Risk assessments also incorporate a variety of professional and science policy judgments (e.g., which models to use, where to locate monitors, which toxicity studies to use as the basis of developing dose-response values). Risk managers therefore need to understand the strengths and the limitations of each assessment, and to communicate this information to all participants and the public.[3] A critical part of the risk characterization process, therefore, is an evaluation of the assumptions, limitations, and uncertainties inherent in the risk assessment in order to place the risk estimates in proper perspective.

One of the key purposes of uncertainty analysis is to provide an understanding of where the estimate of exposure, dose, or risk is likely to fall within the range of possible values. At lower tiers of analysis, this often is expressed as a subjective confidence interval within which there is a high probability that the estimate will fall. When a more quantitative understanding is important to the risk management decision, a related analysis, termed "sensitivity analysis" or "analysis of uncertainty importance," is often performed to identify the relative contribution of the uncertainty in a given parameter value (e.g., emission rate, ingestion rate) or model component to the total uncertainty in the exposure or risk estimate.[15] Often this is used either to identify which parameter values should be varied to provide high-end vs. central-tendency risk estimates, or to identify parameter values where additional data collection (or modeling effort) can increase the confidence in the resulting risk estimate.

There are numerous sources of uncertainties in air toxics risk assessments, and each merits consideration. The degree to which these sources of uncertainty need to be quantified, and the amount of uncertainty that is acceptable, varies considerably on a study-specific basis. For a screening-level (tier 1) analysis, a high degree of uncertainty is often acceptable, provided that conservative assumptions are used to bias potential error toward protecting human health. The use of conservative assumptions is intended to result in a health-protective estimate where the risk assessor is confident that the risk estimate is unlikely to be *greater* than the point estimate of risk. In other words, the point estimate of risk is expected to be at the high end of the range of possible values. The uncertainty characterization for a tier 1 analysis commonly is limited to a qualitative discussion of the major sources of uncertainty and their potential impact on the risk estimate. At higher tiers of analysis, sensitivity analysis to quantify the impact of varying input parameter values (or model algorithms) on the risk estimate, or more complete quantitative uncertainty analysis, commonly are performed to more fully describe the range of possible or plausible values. There are three general approaches for tracking uncertainty through the risk assessment:

- *Qualitative Approach.* This approach involves developing a quantitative or qualitative description of the uncertainty for each parameter and indicating the possible influence of these uncertainties on the final risk estimates given knowledge of the models used.

- *Semi-Quantitative Approach.* This approach involves: (1) using available data to describe the potential range of values that the parameters

> **Sources of Uncertainty**
>
> - **Scenario uncertainty**. Information to fully define exposure or risk is missing or incomplete
> - **Model uncertainty**. Algorithms or assumptions used in models may not adequately represent reality
> - **Parameter uncertainty**. Values for model parameters cannot be estimated precisely
> - **Decision-rule uncertainty**. Policy and other choices made during the risk assessment may influence risk estimates

might assume; (2) performing sensitivity analysis to identify the parameters with the most impact on the risk estimate; and (3) performing sensitivity analysis to compute the range of exposure or risk estimates that result from combinations of minimum and maximum values for some parameters and mid-range values for others.

- *Quantitative Approach.* Probabilistic techniques such as Monte Carlo simulation analysis can explicitly characterize the extent of uncertainty and variability in risk assessment, especially in the exposure assessment step. Using these techniques, important variables in the exposure assessment, as well as in the other parts of the risk assessment, are specified as distributions (rather than as single values) according to what can be expressed about their underlying variability and/or uncertainty. Values are sampled repeatedly from these distributions and combined in the analysis to provide a range of possible outcomes. While this technique can offer a useful summary of complex information, it must be noted that the analysis is only as certain as the underlying data (and assumed forms of the distribution of data values in the population). It is important that the risk assessor clearly expresses individual modeled variables in a way that is consistent with the best information available. Highly quantitative statistical uncertainty analysis is usually not practical or necessary for most air toxics risk assessments. The general quantitative approach to propagating or tracking uncertainty through probabilistic modeling is described in Volume 1, Chapter 31.

Highly quantitative statistical uncertainty analysis will generally be limited to situations where the additional information (i.e., beyond a deterministic risk assessment) has some potential to influence a risk management decision. Probabilistic analyses, from which the risk results can be presented as valid probability distributions, require additional resources.

Additional discussions of uncertainty analysis, including practical approaches to the assessment and presentation of the principal sources of uncertainty in risk assessments are provided in EPA's *Risk Assessment Guidance for Superfund,*[16] guidance documents prepared by EPA and other authors,[4] and in Volume 1 (Chapter 13) of this reference library.

6.0   Tier 1 Inhalation Analysis

6.1   Introduction

This section describes an example approach for performing a Tier 1 inhalation risk assessment. Exhibit 11 provides an overview of this example approach. SCREEN3 or similar models are used for both chronic and acute exposure estimates. Tier 1 analyses incorporate simplified assumptions and default values for facility/source-specific modeling inputs that are not readily available, and allow a simple, health-protective risk estimate to be calculated. The resulting exposure estimates are likely to be higher than actual exposures. If the facility/source passes this screening analysis, the risk manager can be reasonably confident that the likelihood for significant risk is low. This example Tier 1 approach is not intended to prescribe a specific approach that must be used by EPA or others in a particular risk assessment activity. *In particular, various modifications to this tiered approach may be both cost-effective and appropriate, such as adding intermediate-level tiers that incorporate some features of the higher and lower tiers, or conducting iterative, more refined analyses within a given tier. Also, S/L/T agencies may have specific tiered assessment protocols and/or specific modeling guidelines that must be followed. Therefore, consultation with appropriate regulatory agencies is highly recommended.*

Note that this example Tier 1 approach does not incorporate monitoring. While existing monitoring data may be used as inputs to a Tier 1 analysis, if the Tier 1 results are not sufficient for the risk management decision, it may be cost-effective to conduct additional analysis (Tier 2 or Tier 3) before implementing a new monitoring program.

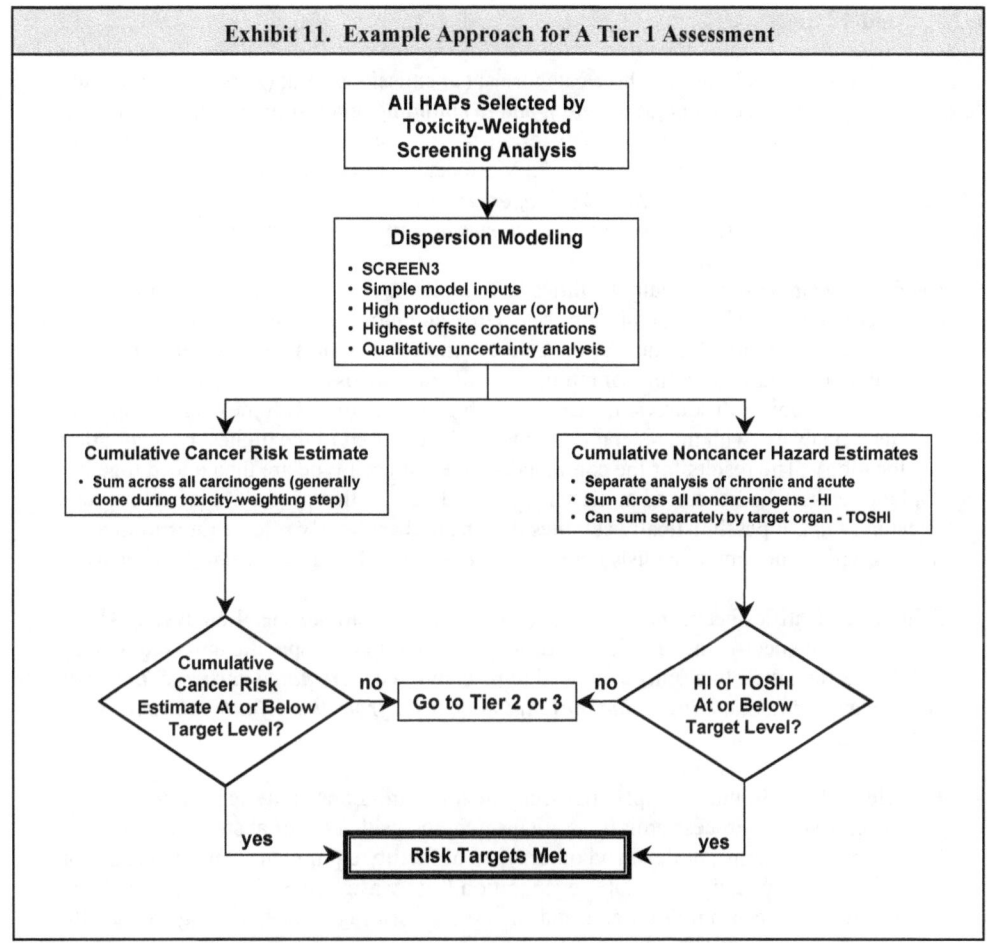

Exhibit 11. Example Approach for A Tier 1 Assessment

All HAPs Selected by
Toxicity-Weighted
Screening Analysis

Dispersion Modeling
- SCREEN3
- Simple model inputs
- High production year (or hour)
- Highest offsite concentrations
- Qualitative uncertainty analysis

Cumulative Cancer Risk Estimate
- Sum across all carcinogens (generally done during toxicity-weighting step)

Cumulative Noncancer Hazard Estimates
- Separate analysis of chronic and acute
- Sum across all noncarcinogens - HI
- Can sum separately by target organ - TOSHI

Cumulative Cancer Risk Estimate At or Below Target Level?

Go to Tier 2 or 3

HI or TOSHI At or Below Target Level?

no ← → no

yes → **Risk Targets Met** ← yes

6.2   Fate and Transport Modeling

SCREEN3 can be used to predict maximum hourly ambient concentrations on the centerline of a plume downwind from a source for all HAPs selected for Tier 1 during the toxicity-weighted screening analysis (see Chapter II). The hourly concentrations are used directly for acute exposure estimates and converted to maximum annual concentrations using a fixed screening factor for chronic exposure estimates (see below). SCREEN3 uses generalized meteorological data, some facility/source-specific information (e.g., terrain and building dimension information), and facility/source-specific emissions data to estimate downwind ambient air concentrations of HAPs within a user-specified radius from the source up to 50 km (30.8 miles). Estimates of excess cumulative cancer risk and noncancer hazard can be calculated separately from model outputs, or can be calculated by the model if emission rates are input in toxicity-weighted form. Risk estimates are based on the concentrations at the point of maximum concentration.

6.2.1   Model Inputs

Input data for SCREEN3 fall into three categories: (1) emissions data; (2) release parameters (e.g., source type, release temperature, etc.); and (3) building and terrain considerations (e.g., building downwash and simple or complex terrain). Note that several default parameters for meteorological data are built into SCREEN3. More detailed information is provided in the SCREEN3 User's Guide.[8] *Note that S/L/T agencies may have emissions and other source information in documents such as state inventories and inspection reports.*

- **General considerations**. Many facilities have multiple emission sources (each with its characteristic release height, emissions rate, etc.), and more than one type of emission source within a single facility. Because SCREEN3 can model a single source per model run, and is not able to aggregate the results of multiple model runs across different sources, it is most efficient to combine all sources of each source type (e.g., point, volume) into a single source for Tier 1 analysis (with the receptor assumed to be exposed to the highest concentration at any location). The results for the combined sources of each type are then added together, as if all the releases had occurred at the same point. Because this procedure is health-protective screening (e.g., it pessimistically assumes that the highest hourly release rates from all sources will occur simultaneously), it is appropriate for a Tier 1, screening-level analysis.

    Exhibit 12 identifies health-protective default values to facilitate Tier 1 analyses. All facility/source-specific information needs to be provided in the specific units required by SCREEN3 (and therefore units may need to be converted). *As noted earlier, S/L/T agencies may have specific modeling guidelines that may differ from the default values presented in Exhibit 12.*

- **Emissions data**. In this example approach, the total emissions is the sum of *process* and *fugitive* emissions. Process emissions are discrete losses that occur at stacks or process vents from reactors, columns, boilers, and other types of facility equipment. Fugitive emissions result from facility equipment leaks, evaporation from waste products, losses from the raw material feed, losses from in-process and final process storage tanks, loading and handling losses, and losses from other non-discrete sources.

The risk assessor is encouraged to use the best emissions data available for the facility, including chemical speciation information. If speciated emissions data are not readily available, an assumption that all HAPs are emitted in the most toxic form commonly is used in Tier 1 assessments to ensure that health-protective risk estimates are produced.

> **Where to Find Source Emissions Information**
>
> - Facility records or purpose-specific data collection
> - National Emissions Inventory
> - State and Federal risk assessment reports
> - Background Information Documents that support Maximum Achievable Control Technology Standards (MACT), available from EPA's MACT docket
> - State approved emission permits
> - Toxics Release Inventory

Exhibit 12. Example Default Values for SCREEN3 Tier 1 Inputs[a]		
Input Parameter	**Stack/Vent Health-protective Default**	**Fugitive Source Health-protective Default**
Source type	Point	Area
Emission rate (g/s)[b]	Actual	Actual (g/s/m^2)
Stack height (m)	5	1
Stack diameter (m)	0.1524	n/a
Stack temperature (degrees Kelvin)	293	n/a
Stack gas exit velocity (m/s)	1	n/a
Ambient temperature (degrees Kelvin)	293	n/a
Length of larger side of area source (m)	n/a	5
Length of smaller side of area source (m)	n/a	5
Receptor height above ground (m)	0	0
Urban/rural	Rural	Rural
Building downwash (Yes/No)	Yes	Yes
Building height (m)	5	5
Minimum horizontal width (m)	30	30
Maximum horizontal width (m)	30	30
Use simple terrain? (Yes/No)	Yes	Yes
Use complex terrain? (Yes/No)	No	No
Use simple elevated terrain? (Yes/No)	No	No
Meteorology	1-Full	1-Full
Automated distance array? (Yes/No)	Yes	Yes
Minimum and maximum distances (m)	100; 5,000	100; 5,000
Discrete distances (m)	No	No
Printout results	Yes	Yes

[a] *S/L/T agencies may have specific modeling guidelines that may differ from these default values*
[b] Use annual average for chronic exposure estimates and 1-hour maximum for acute exposure estimates

One common Tier 1 approach is to use emissions data that are representative of maximum annual emissions (e.g., a high-production year). For **chronic exposures**, the approach uses the average hourly emissions rate for the high-production year. For **acute exposures**, the approach uses the greater of (a) the maximum hourly rate, or (b) ten times the average hourly rate. The ratio between a longer-term maximum concentration and a 1-hour maximum will depend upon the duration of the longer averaging time, source characteristics, local climatology and topography, and the meteorological conditions associated with the 1-hour maximum. Because of the many ways in which such factors interact, it is not practical to categorize all situations that will typically result in any specified ratio between the longer-term and 1-hour maxima. EPA's *Screening Procedures for Estimating the Air Quality Impact of Stationary Sources, Revised*[6] identifies ratios for a "general case," and the risk assessor is given some flexibility to adjust those ratios to represent more closely any particular point source application where actual meteorological data are used. Emissions of multiple HAPs from a single source can be combined into a single emission using toxicity weighting (see Section 6.2.2 below).

To obtain the estimated maximum concentration for a 3-, 8-, 24-hour or annual averaging time, multiply the 1-hour maximum concentration by the indicated factor in the table below. The numbers in parentheses are recommended limits to which one may diverge from the multiplying factors representing the general case. For example, if aerodynamic downwash or terrain is a problem at the facility, or if the emission height is very low, it may be necessary to increase the factors (within the limits specified in parentheses). On the other hand, if the stack is relatively tall and there are no terrain or downwash problems, it may be appropriate to decrease the factors. If the risk assessment involves a regulatory action, the risk assessor is encouraged to discuss these factors with the appropriate regulatory authorities.

Averaging Time	Multiplying Factor
3 hours	0.9 (±0.1)
8 hours	0.7 (±0.2)
24 hours	0.4 (±0.2)
Annual	0.08 (±0.02)

Note that some assessments may be based on permit limits (e.g., if their purpose is to determine if those limits provide adequate protection).

The State of California has determined that the conversion multiplying factors noted above may not be biased toward over-prediction when the source is *not* a continuous release. Appendix H of the *Air Toxics Hot Spots Program Guidance Manual for Preparation of Health Risk Assessments* (August 2003; http://www.oehha.ca.gov/air/hot_spots/pdf/HRAguidefinal.pdf) presents approaches for developing multiplying factors for (a) non-standard averaging periods with a continuous release; (b) intermittent releases (e.g., releases that occur only from 8 AM to 6 PM); (c) systematic releases (e.g., a "clean-out" release once per day); and (d) random releases.

- **Release type**. SCREEN3 will prompt the user for source-specific characteristics. Point sources are localized releases from stacks or vents. Flare sources are point sources that have a high release rate (e.g., flames). Area sources are emissions that are spread over an area (e.g., a landfill or lagoon). Volume sources are three-dimensional releases (e.g., fugitive leaks from an industrial facility). When modeling a facility with multiple sources, assessors typically model each source separately and sum the resulting maximum concentrations (see below).

- **Physical release parameters**. SCREEN3 will prompt the user for the release parameters appropriate to the type of release. For example, SCREEN3 requires stack height, stack diameter, stack gas exit velocity, and stack temperature in order to model a point source. When available, it is preferable to use site- or source-specific release parameters. Example default values for point, area, and volume sources (Exhibit 12) can be used *when facility/source-specific values are not available.*

- **Meteorological data and building and terrain considerations.** For this example approach, the following inputs are selected: (1) the "actual" urban/rural dispersion coefficient (based on the site setting); (2) "full meteorology" (which considers the effects of all atmospheric stability classes); (3) building downwash (using either the actual dimensions of the largest building in the facility or the default dimensions in Exhibit 12); and (4) "simple terrain" (elevated terrain can be entered if appropriate

> **Building Downwash**
>
> Volume 2 of the ISC User's Guide (Section 1.1.5.3) provides a rule-of-thumb that a building is considered sufficiently close to a stack to cause wake effects when the distance between the stack and the nearest part of the building is less than or equal to five times the lesser of the height or the projected width of the building. This relationship is obviously much more complicated with complex building shapes. The ISC User's guide is available on EPA's SCRAM website (http://www.epa.gov/ttn/scram/).

for the facility/source being evaluated; otherwise, terrain should be assumed flat). Note that SCREEN3 uses built-in meteorological data (see Volume 1, Chapter 9).

6.2.2 Model Runs

SCREEN3 provides estimates of maximum one-hour ambient concentrations. *If multiple model runs are performed (e.g., separate runs for point sources and area sources), a conservative approach is to sum the maximum predicted model results from each run across all sources.* This approach assumes that the maximum impact for each source occurs at the same location and the same hour. These maximum one-hour concentrations can be used to evaluate acute noncancer hazard as described in Section 6.5 below.

SCREEN3 models one HAP for each run of the model. For multiple chemicals, the modeler can input 1 gram/year as the emission rate and run the model once (or once for each source type). Ambient concentrations for each HAP can then be calculated by multiplying the model results by the annual emission rate (in tons) for each HAP.

6.3   Exposure Assessment

Tier 1 assessments will generally assume that the highest ambient concentrations predicted by SCREEN3 are equivalent to exposure concentrations. As noted above, for chronic exposures, the approach uses the average hourly emissions rate for the high-production year. For acute exposures, the approach uses the maximum hourly rate, or 10 times the average hourly rate, whichever is greater. If separate model runs were done for some sources (e.g., to separate point sources from area sources), this example approach sums the maximum exposure concentrations from each model run before characterizing risk. This procedure effectively treats all the releases as occurring in the same place.

6.4   Risk Characterization

Risk characterization for Tier 1 is limited to estimating inhalation risk at the point of maximum concentration (calculated separately for cancer risk, chronic noncancer hazard, and acute noncancer hazard). The exposure concentrations described in Section 6.3 are used to calculate risk and hazard according to basic equations presented in Section 5 above. Background concentrations are often not explicitly considered in Tier 1.

- **Cancer risk**. Excess cumulative cancer risk is calculated by multiplying the maximum annual air concentration by the inhalation IUR for each HAP and summing across all HAPs.

- **Chronic noncancer hazard**. Chronic cumulative noncancer hazard (the Hazard Index, or HI) is calculated by dividing the maximum annual air concentration by the RfC for each HAP and summing across all HAPs. If the resulting HI exceeds the noncancer target HI level of regulatory interest, the HI results can be subdivided into a separate Target Organ Specific Hazard Index (TOSHI) for each target organ of concern (see Volume 1, Chapter 13 for a more detailed discussion on how to calculate a TOSHI.).

- **Acute noncancer hazard**. As discussed in Volume 1, Chapter 13, available acute dose-response values are more diverse than chronic values, because they were developed for different purposes and considering different exposure durations. The most effective characterization of acute risk often is to compare the maximum estimated hourly concentrations with a range of acute dose-response values from sources described in Section 4 (see an example comparison in Exhibit 13). If the ambient concentration is lower than all the acute benchmarks, it is generally reasonable to conclude that the potential for significant acute risk is low. If the concentration exceeds some benchmarks but not others, the assessment should describe the benchmark (e.g., is it a no-effect level, or a concentration at which mild effects occur?) and discuss the implications of exceeding it.

6.4.1   Reporting Results

A relatively simple summary can be used to report results but should still ensure the results are both transparent and reproducible. Examples of reports prepared for EPA's purposes can be found in EPA's residual risk "test memos;" these may or may not be appropriate examples for the specific purposes of other risk assessments. The summary generally will include the following information:

- Documentation of input parameters, output spreadsheets, and risk characterization, with special emphasis on comparing estimated risks to risk targets;

- A simple presentation describing the assessment's purpose (e.g., to determine whether risk is below levels of concern) and the outcome relative to that purpose (e.g., low risk is not demonstrated).

- Documentation of anything in the analysis that is discretionary (i.e., anything that is facility-specific), such as emissions characteristics or choice of a meteorological station other than the nearest.

Exhibit 13. Example Presentation of Risk Characterization Results for Acute Exposure

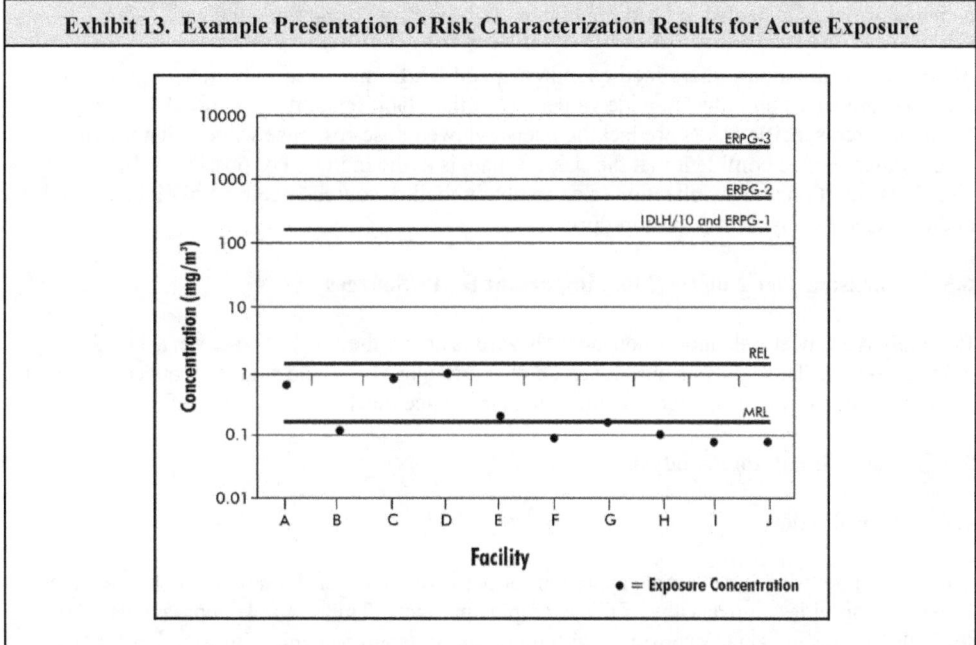

This example illustrates one way to graphically compare estimated exposure concentrations (ECs) from multiple facilities (or sources within a facility) to available acute dose-response values. In this example, estimated ECs from facility A, C, D, E, and G were at or above MRLs but below all other acute values. The risk characterization would need to clearly describe the significance of these results in terms of potential health effects and related uncertainties.

6.4.2 Assessment and Presentation of Uncertainty

Risk managers need to understand the strengths and the limitations of the Tier 1 assessment. A critical part of the risk characterization process, therefore, is an evaluation of the assumptions, limitations, and uncertainties inherent in the Tier 1 risk assessment in order to place the risk estimates in proper perspective.[3] Tier 1 risk assessments will typically include a quantitative or qualitative description of the uncertainty for each parameter and indicating the possible influence of these uncertainties on the final risk estimates given knowledge of the models used (i.e., the qualitative approach discussed in Section 5.4 above). The use of conservative assumptions in Tier 1 is intended to result in result in a situation where the risk assessor is confident that the risk estimate is unlikely to be *greater* than the point estimate of risk. In other words, the point estimate of risk is expected to be at the high end of the range of possible values.

There may be situations where facility/source-specific information is so limited (e.g., emissions estimates are based on industry-wide values rather than facility/source-specific data; many facility/source-specific HAPs are lacking peer-reviewed dose-response values) that the risk assessor may not be confident that the risk estimate is at the high end of possible values. In these situations, additional data collection (e.g., to obtain facility/source-specific emissions estimates) and/or a Tier 2 analysis may be appropriate.

6.5 Focusing Tier 2 on the Most Important HAPs/Sources

If cumulative cancer risk and/or noncancer hazard is above the level of concern, a Tier 2 analysis may be needed. This analysis may focus on the particular HAPs and sources that account for a high proportion of the estimated risk (at Tier 1) to reduce the Tier 2 analytical effort.

7.0 Tier 2 Inhalation Analysis

7.1 Introduction

This section describes an example approach for performing a Tier 2 inhalation risk assessment. Exhibit 14 provides an overview of this example approach. This example approach uses HEM-3 for both chronic and acute exposures, with more facility/source-specific information for HAPs, sources, and potentially exposed populations. *Tier 2 analyses may often be limited to those HAPs and sources that collectively account for a high proportion of the potential risk or hazard that was estimated in Tier 1. If the facility/source database permits, risk assessors may be able to skip Tier 1 entirely and start with Tier 2.* For this example Tier 2 approach, HEM-3 provides an estimate of ambient concentrations from all the facility's sources combined at multiple Census block internal points, including the Census block with the highest ambient concentration. In this approach to a Tier 2 assessment, these Census block concentrations are used as surrogates for exposure concentrations for the people residing there. Note that more complex/advanced exposure modeling is used in the example Tier 3 approach described in Section 7.

The most important differences from Tier 1 are: (1) the exposure surrogate shifts from the maximum offsite ambient concentration to the concentrations within Census blocks (i.e., where people actually reside); (2) individual modeling of release points is performed; and (3) average production year data are used for cancer risk estimates, while high-production year data are used for noncancer hazard estimates.

Note that this example Tier 2 analysis focuses on current land use scenarios. It may be appropriate to examine future land use scenarios.

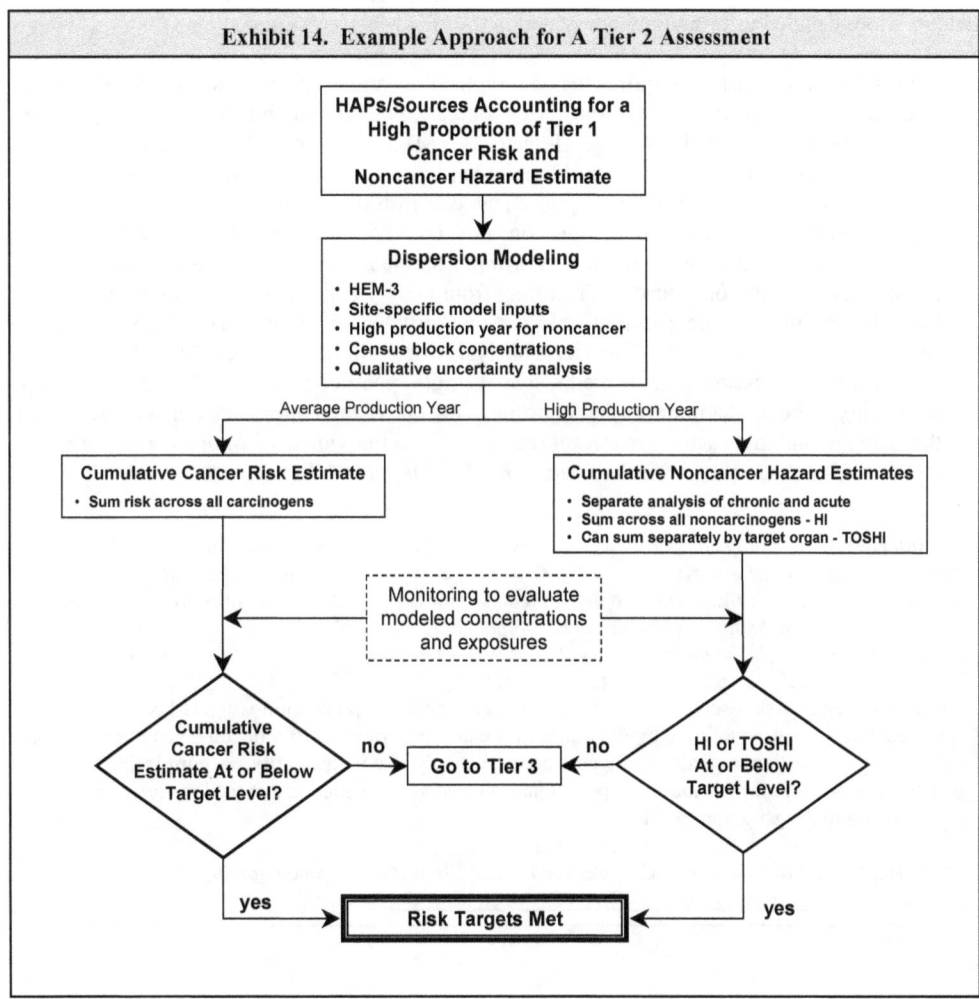

Exhibit 14. Example Approach for A Tier 2 Assessment

7.2   Fate and Transport Modeling

7.2.1   Model Inputs

Input data for HEM-3 falls into six categories: (1) source location; (2) emissions data; (3) stack/vent parameter data; (4) pollutant specific data (reactivity and dose-response data); (5) meteorological data; and (6) population data. Refinements for Tier 2 are described briefly below. More detailed information is provided in the HEM-3 User's Guide.[7]

- **Source location**. HEM-3 requires the geographical location (latitude and longitude) of each source being simulated. Some of the above sources of geographic information may provide coordinates in Universal Transverse Mercator (UTM) units. The HEM-3 contains a program to convert UTM data into latitude and longitude (e.g., as provided by the use of a GPS receiver). In many instances there is more than one emission source (each with its characteristic release type, height, emissions rate, etc.) within a single facility. HEM-3 can model each emission source individually within a single model run. HEM-3 also computes the total ambient air concentrations resulting from facility emissions by summing the individual estimates. The geographical location (latitude and longitude) of all sources being simulated must be identified. The Tier 2 analysis often will not include a default option to assume that all emission sources are located at single specified latitude and longitude within the facility. The model requires that coordinate data be obtained for each emission source in the analysis, and that each emission source is modeled individually. *In other words, the location of each source within the facility should be input to HEM-3.*

> The **internal point** is only an approximation of where people live. The Census Bureau[a] defines the internal point as *a set of geographic coordinates (latitude and longitude) that is located within a specified geographic entity. A single point is identified for each entity; for many entities, this point represents the approximate geographic center of that entity. If the shape of the entity causes this point to be located outside the boundary of the entity or in a water body, it is relocated to land area within the entity.* The term "centroid" (which is used interchangeably with internal point) is the term more commonly used by risk assessors when referring to geographic or population-weighted centers. Note that the internal point established by the Census Bureau is generally set to reflect the geographic center of the entity in question, regardless of where people actually live in that entity. A **population weighted internal point** is a modified point that is considered to better reflect the location of the people living in the geographic entity.
>
> [a]U.S. Department of Commerce, U.S. Census Bureau. 2000. *Geographic Glossary (Census 2000).* Available at: http://www.census.gov/geo/www/tiger/glossry2.pdf

- **Emissions data**. Depending on the source, the emissions being modeled may be process or fugitive emissions. The risk assessor should use the best site-specific emissions data available, including chemical speciation. As with Tier 1, HEM-3 requires emissions data in the form of an annual emissions rate (e.g., tons/year). HEM-3 can be used to assess both chronic and acute exposures. To do this, select "yes" for the "include hourly emissions variations" choice, and provide a file of hourly emission rate factors (e.g., 10 times the annual average) appropriate to the type of source for each HAP. Emission rate factors based on measured data are preferred, but if measurements are not available, engineering judgment based on the nature of the process can be used. For infrequent releases (e.g., less than once a month), the assessor may assume that the total annual emission rate is released during one episode and use engineering judgment to determine the length of the episode.

- **Release parameter data**. Release parameter data are necessary for dispersion modeling of the chemical emissions. Facility/source-specific data generally are preferred for a Tier 2 analysis. When these are not available, surrogate data often can be found in the same data sources as the emissions data. HEM-3 release parameter data requirements include:
  - Stack or area source release height and diameter (meters);
  - Dimensions of area source (meters);
  - Stack diameter (meters);
  - Gas discharge temperature (degrees Kelvin);
  - Gas emission (exit) velocity (meters/second);
  - Height of the initial plume (meters); and
  - Building dimensions (meters, if relevant).

- **Atmospheric reactivity and deposition of chemicals**. Because the highest inhalation exposures are generally near the source, and little time passes between the release and the exposures, atmospheric reactivity and deposition of fine particles generally do not need to be considered in Tier 2.

- **Meteorology**. HEM-3 uses meteorological data measured both at the surface and in the upper atmosphere in its dispersion calculations. Surface data used by HEM-3 are pre-processed hourly meteorological observations from a user library, or information from a user-specified meteorological station. Upper atmosphere data are twice-daily mixing height values. User-supplied data sets need to be pre-processed by the EPA RAMMET to fit the HEM-3 input format. Surface and upper atmosphere data sets are available from the National Climate Data Center in Asheville, NC or from the EPA SCRAM website (www.epa.gov/scram001/index.htm).

- **Population**. In this example of a Tier 2 analysis, modeling is focused on the *locations where people actually live*. HEM-3 automatically identifies the exposed population using the Census block internal points already included in the model. These locations are modeled directly within a user-specified radius and are interpolated from locations on the polar coordinate receptor grid for Census blocks outside the radius (Exhibit 15). For Tier 2 analyses, all people residing in a Census block are assumed to be continuously exposed to the annual average concentration (and exposed for one hour to the maximum hourly concentration) at the interior location within the Census block. Unless otherwise specified, HEM-3 calculates concentrations for each Census block within 50 kilometers of the source's location. If a Census block is exposed to emissions from more than one source, HEM-3 sums the impacts from each source for each Census block. The model outputs include the risks posed by each HAP at each Census block internal point, the risks from all HAPs combined, and the numbers of people living in Census blocks whose ambient levels fall within specified risk ranges.

Exhibit 15. Example of Interpolation to Calculate Concentrations at Census Block Internal Points

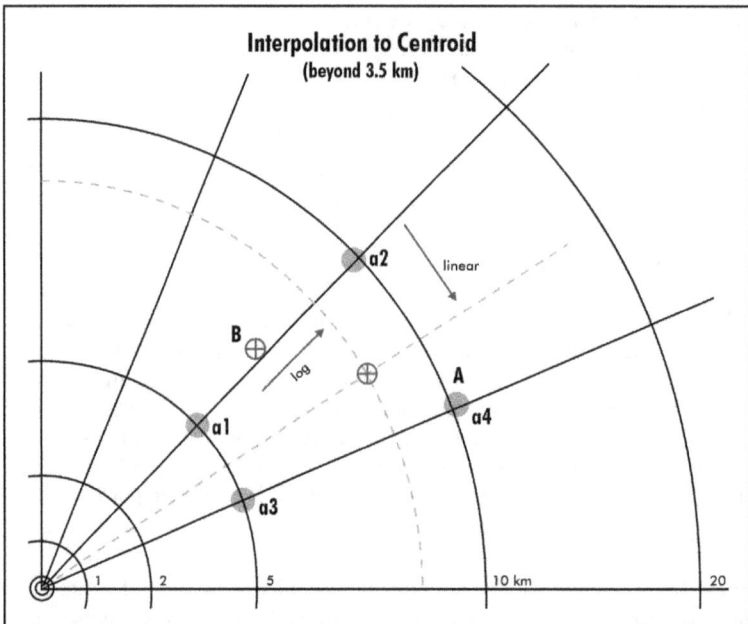

Interpolation to Centroid
(beyond 3.5 km)

For polar grids, a two-step interpolation is used, starting with the modeled concentrations at the nearest locations (e.g., a1, a2, a3, and a4 in the graph above). The first interpolation is in the radial direction (i.e., along the two adjacent radial lines [a1,a2] and [a3, a4] in the graph). The concentration is estimated at the intersection of each radial line with the concentric circle that intersects the receptor location (i.e., at the same radial distance from the source as the internal point). This interpolation is performed under the assumption that the logarithm of the concentration decreases in proportion to the increase in the logarithm of the distance from the source (i.e., a log-log interpolation). The second interpolation is in the azimuthal direction (i.e., along the concentric circle that intersects the internal point). This interpolation is performed under the assumption that the change in concentration is proportional to the distance around the circle between the two radial lines (i.e., linear interpolation).

7.2.2 Model Runs

The following input values can be used to set up HEM-3 model runs for a Tier 2 analysis:

- On screen 1, select "inhalation," "yes" to both annual and acute concentrations, "1-hour" for the length of acute exposure, and "rural" for the dispersion environment (unless the facility/source is located in a densely-populated area).

- Provide the names of the emissions file and the emissions location file, which should be in the correct spreadsheet format (HEM-3 provides format guidelines).

- Select "no" for the deposition and depletion choices.

- Select "yes" for the terrain elevation choice if the facility/source is located in an area with substantial elevation changes; otherwise select "no" for flat topography.

- Select "no" to providing alternate IURs and RfCs.

- Select dimensions of the area to be modeled that are appropriate to the facility. The maximum radius is 50 kilometers, but for most facilities a radius of 30 kilometers will capture virtually all important exposures. Select 0 as the minimum radius.

- Select the "distance within which blocks are modeled individually" to capture the Census blocks with the highest concentrations.

- Choose the appropriate HEM-3 meteorological data file, or provide a facility/source-specific, pre-processed file. Sources of meteorological data are provided in Volume 1, Appendix G.

7.3   Exposure Assessment

The Tier 2 assessment approach described here uses predicted ambient concentrations as surrogates for exposure concentrations. HEM-3 provides estimates of annual average concentrations within Census blocks and combines them with chronic dose-response values for cancer and noncancer effects to estimate risk.

7.3.1   Chronic Exposures

HEM-3 provides ambient concentrations for the Census block with the highest annual average levels, by individual HAP and emission point. HEM-3 also calculates the upper-bound lifetime cancer risk and noncancer HQs associated with continuous exposure to those ambient concentrations, and it sums cancer risk across all HAPs and noncancer HQs by target organ. These results can be presented in summary tables (e.g., Exhibits 16 to 19, below).

7.3.2 Acute Exposures

HEM-3 also provides estimates of the highest average ambient concentration predicted anywhere within the assessment area. To estimate the maximum 1-hour ambient concentration, this example approach multiplies the average concentration by the same screening factor used in Tier 1 (discussed in section 7.2, above, e.g., 10-fold). These maximum 1-hour concentration estimates for each HAP are used as surrogates for acute exposures, to characterize acute noncancer hazard as described in Section 7.5. *Note that the maximum hourly concentrations for different HAPs may occur in different locations; however HEM-3 automatically sums across pollutants, thereby addressing this issue.*

7.3.3 HEM-3 Outputs

HEM-3 produces six different output tables that make the following presentations of ambient concentrations and risks:

- The maximum individual risk table contains the upper bound lifetime cancer risk, total hazard index, and target organ-specific hazard indexes for the Census block with the highest ambient concentrations (including Census ID and location information).

- The maximum offsite impacts table provides the same information as the maximum individual risk table, except the location is the offsite point of highest ambient concentration, whether populated or not.

- The cancer risk/exposure table shows the numbers of people living in Census blocks whose annual mean concentrations correspond to specified risk ranges (e.g., greater than or equal to 1 in 1 million, 10 in 1 million).

- The noncancer risk/exposure table provides similar information regarding numbers of people living in Census blocks whose concentrations correspond to specified TOSHI ranges.

- The risk breakdown table provides the incremental contribution to individual cancer risk and hazard quotient in the most exposed Census block by each HAP from each source.

- The incidence table presents the incremental contribution to total estimated cancer incidence within the modeled duration by each HAP from each source.

These presentations overlap substantially, and often may be condensed, as for the four example summary tables in Exhibits 16 to 19 below.

Exhibits 16 and 17 illustrate estimates of individual cancer risk and noncancer hazard to a typical resident of the most exposed Census block, using the modeled ambient concentration within the block as a surrogate for exposure concentration. These tables are sorted to show the contribution of each source in the assessment, and the combination of all sources. Note that the highest exposure associated with different sources, and the total of all sources, may occur in different Census blocks. Therefore, the subtotals in these tables may not be additive across sources.

Exhibits 18 and 19 illustrate estimates of the aggregate cancer risk and noncancer hazard to all Census block residents within the modeling domain. Receptor populations have been placed in risk categories, and are shown as the numbers of people who live in Census blocks where the modeled ambient concentration (if inhaled continuously), corresponds to specific levels of risk or hazard. For example, the assessment predicts Census blocks where 36,599 people reside have ambient concentrations equivalent to upper-bound lifetime cancer risk levels between 1 and 10 in 1 million. For characterizing acute risks, HEM-3 calculates the maximum average annual offsite concentrations. These can be adjusted by the appropriate screening factors, and graphically compared with acute benchmarks (as illustrated earlier in Exhibit 13).

7.3.4 Monitoring Data

It may be helpful to conduct a monitoring program to evaluate or further characterize exposure concentrations at key locations (e.g., in the Census blocks where HEM-3 indicates relatively high exposures). Volume 1 (Chapter 10) provides an overview of available approaches for assessing air quality via monitoring.

7.4 Risk Characterization

Risk characterization for this example Tier 2 assessment focuses on inhalation risk to a receptor at the point of maximum concentration (calculated separately for cancer risk, chronic noncancer hazard, and acute noncancer hazard). This example approach uses the exposure concentrations noted in Section 7.4 to calculate risk and hazard according to basic equations presented in Section 5 above. *Note that the maximum concentrations for cancer risk, chronic noncancer hazard, and acute noncancer hazard may occur at different locations.*

Although background concentrations (i.e., ambient levels originating from sources other than the facility/source) are not explicitly considered in this example Tier 2 approach, situations may exist where background is a potential concern. In such cases, monitoring data and/or modeled results from community-wide assessments can be used to support a risk assessment for background concentrations, and for the comparison of risk/hazard associated with facility/source emissions with that associated with background concentrations.

Exhibit 16. Example Presentation of HEM-3 Individual Cancer Risk Estimates					
Receptor: Individual at Census Block Internal Point with Highest Cancer Risk Estimate					
Upper-bound Lifetime Cancer Risk		Emission Type	Risk	Concentration ($\mu g/m^3$)	Emissions (tons/yr)
Source	Pollutant				
1	1,3-Butadiene	volume	8e-07	0.03	20
1	Arsenic compounds	point	8.E-07	0.0002	0.1
1	Benzene	volume	8e-08	0.01	7
1	Cadmium compounds	point	3.E-07	0.0001	0.1
1	Chromium (VI) compounds	point	6.E-09	0	0.0003
1	Polycyclic Organic Matter	volume	2.E-09	0.00002	0.01
1 Total			2.E-06		
2	Arsenic compounds	point	5.E-08	0.00001	0.0005
2	Cadmium compounds	point	2.E-08	0.00001	0.0005
2	Chromium (VI) compounds	point	1.E-09	0	0
2 Total			7.E-08		
3	Arsenic compounds	point	8.E-08	0.00002	0.0008
3	Cadmium compounds	point	3.E-08	0.00002	0.0008
3	Chromium (VI) compounds	point	2.E-09	0	0
3 Total			1.E-07		
All	1,3-Butadiene	combined	8.E-07	0.03	
All	Arsenic compounds	combined	2.E-06	0.004	
All	Benzene	combined	8.E-08	0.01	
All	Cadmium compounds	combined	6e-07	0.0004	
All	Chromium (VI) compounds	combined	5.E-08	0	
All	Polycyclic Organic Matter	combined	1.E-11	0	
All Total			0		

[a] Standard rules for rounding apply which will commonly lead to an answer of one significant figure in both risk and hazard estimates. For presentation purposes, hazard quotients (and hazard indices) and cancer risk estimates are usually reported as one significant figure.

Exhibit 17. Example Presentation of HEM-3 Individual Noncancer TOSHI Estimates					
Receptor: Individual at Census Block Internal Point with Highest Noncancer Hazard Estimate					
Respiratory System TOSHI		**Emission Type**	**HQ**[a]	**Concentration ($\mu g/m^3$)**	**Emissions (tons/yr)**
Source	**Pollutant**				
1	Acrolein	volume	3.E-02	0.0007	0.4
1	Arsenic compounds	point	6.E-03	0.0002	0.1
1	Cadmium compounds	point	7.E-03	0.0002	0.001
1	Chromium (VI) compounds	point	5.E-06	0	0.0003
1	Manganese compounds	point	9.E-05	0.000005	0.003
1 Total			5.E-02		
2	Arsenic compounds	point	4.E-04	0.00001	0.0005
2	Cadmium compounds	point	5.E-04	0.00001	0.0004
2	Chromium (VI) compounds	point	1.E-06	0	0
2	Manganese compounds	point	3.E-03	0.0001	0.005
2 Total			3.E-03		
3	Arsenic compounds	point	6.E-04	0.00002	0.0008
3	Cadmium compounds	point	9.E-04	0.00002	0.0008
3	Chromium (VI) compounds	point	2.E-06	0	0
3	Manganese compounds	point	4.E-03	0.0002	0.009
3 Total			0.01		
All	Acrolein	combined	0.03	0.0007	
All	Arsenic compounds	combined	1.E-02	0.0004	
All	Cadmium compounds	combined	0.02	0.0004	
All	Chromium (VI) compounds	combined	4.E-05	0.00004	
All	Manganese compounds	combined	4.E-02	0.002	
All Total			0.1		

[a] Standard rules for rounding apply which will commonly lead to an answer of one significant figure in both risk and hazard estimates. For presentation purposes, hazard quotients (and hazard indices) and cancer risk estimates are usually reported as one significant figure.

Exhibit 18. Example Presentation of HEM-3 Population Cancer Risk Estimates	
Metric: Number of People Living in Census Blocks Corresponding to Specific Estimates of Cancer Risk	
Upper-bound Estimate of Lifetime Cancer Risk	**Population Living in Census Block**
Greater than or equal to 1 in 1,000	0
Greater than or equal to 1 in 10,000	0
Greater than or equal to 1 in 20,000	0
Greater than or equal to 1 in 100,000	0
Greater than or equal to 1 in 100,000	36599
Greater than or equal to 1 in 1,000,000	1102010

Exhibit 19. Example Presentation of HEM-3 Population Noncancer TOSHI Estimates			
Metric: Number of People Living in Census Blocks Corresponding to Specific Estimates of Noncancer Hazard (TOSHI)			
Estimate of Hazard Index	**Population Living in Census Block**		
	Total HI	**Respiratory HI**	**CNS HI**
Greater than or equal to 100	0	0	0
Greater than or equal to 50	0	0	0
Greater than or equal to 10	0	0	0
Greater than or equal to 1.0	0	0	0
Greater than or equal to 0.5	3163	3163	0
Greater than or equal to 0.2	100982	100982	0

7.4.1 Reporting Results

A relatively simple summary can be used to report results, consistent with the need to make the results both transparent and reproducible. Examples of reports prepared for EPA's purposes can be found in EPA's residual risk "test memos;" these may or may not be appropriate examples for the specific purposes of other risk assessments.[d] The summary generally will include the following information:

- Documentation of input parameters, output spreadsheets, and risk characterization, with special emphasis on comparing estimated risks to risk targets;

- A simple presentation describing the assessment's purpose (e.g., to determine whether risk is below levels of concern) and the outcome relative to that purpose (e.g., low risk is not demonstrated); and

- Documentation of anything in the analysis that is discretionary, such as facility-specific emissions characteristics or choice of meteorological station. In particular, the rationale used to select specific receptor locations and methods or calculations used to identify the location associated with the highest individual risk need to be documented.

7.4.2 Assessment and Presentation of Uncertainty

Risk managers need to understand the strengths and the limitations of the Tier 2 assessment. A critical part of the risk characterization process, therefore, is an evaluation of the assumptions, limitations, and uncertainties inherent in the Tier 2 risk assessment in order to place the risk estimates in proper perspective.[3] Tier 2 risk assessments commonly include a quantitative or qualitative description of the uncertainty for each parameter and indicate the possible influence of these uncertainties on the final risk estimates given knowledge of the models used. Tier 2 assessments also may include a semi-quantitative sensitivity analyses. These approaches are described in Section 5.4 above. Sensitivity analyses are discussed in more detail in Volume 1 (Chapters 3 and 13) of this reference library.

7.5 Focusing Tier 3 on the Most Important HAPs/Sources

If cumulative cancer risk and/or noncancer hazard is above the level of concern, a Tier 3 analysis may be needed. This analysis may include all the sources and HAPs assessed in Tier 1, or may be reduced to only those HAPs and sources of primary concern.

---

[d]The "test memos" for a particular source category can be found on the NESHAPs page of EPA's air toxics web site (http://www.epa.gov/ttn/atw/mactfnlalph.html) by selecting the source category and going to the residual risk section of that page.

8.0   Tier 3 Inhalation Analysis

8.1   Introduction

This section describes an example approach for performing a Tier 3 inhalation risk assessment. Exhibit 20 provides an overview of this example approach. *Tier 3 analyses are generally limited to those HAPs and sources that collectively account for a large proportion of the potential risk or hazard estimated in Tier 2.* The example Tier 3 assessment is significantly different than the example Tier 1 and 2 approaches, in that it involves more complex dispersion modeling (e.g., with ISCST3 or AERMOD), specific consideration of population locations, the use of an exposure model (TRIM.Expo$_{Inhalation}$) to account for receptor behavior (e.g., time spent in different microenvironments), and use of a risk model (TRIM.Risk$_{HH}$) to provide a more refined characterization of risk, including multiple estimates of risk (e.g., central tendency and high end).

The Tier 3 analysis often involves considerable flexibility in analytical approach and detail. This example includes the use of:

* A dispersion model (ISCST3 or AERMOD) to calculate hourly and annual average concentrations at user-specified exposure locations. ISCST3 allows considerable spatial refinement in selecting exposure locations, and provides concentrations for user-specified (actual) locations.

* An exposure model (TRIM.Expo$_{Inhalation}$) to combine air concentration information with human activity patterns to develop more precise, population- and site-specific exposure estimates. The user specifies the geographic area to be modeled, the number of individuals to be simulated to represent the study area population, and the demographic unit of resolution for the outputs. The model produces exposure estimates specific to modeled individual and demographic unit (e.g., census tract or block, etc). Because the modeled individuals represent a random sample of the population of interest, the distribution of modeled individual exposures can be extrapolated to the larger population.

* A risk model (TRIM.Risk$_{HH}$) to calculate cumulative excess cancer risk and noncancer hazard associated with the modeled exposures. TRIM.Risk$_{HH}$ calculates human health risk metrics, documents model inputs and assumptions, and displays results.

* Monitoring data may be used to evaluate or further characterize exposure concentration and exposure estimates.

The Tier 3 analysis may incorporate probabilistic inputs and model outputs using the TRIM modeling system. If a potential regulatory action is involved, the scope, method, and inputs for probabilistic modeling should be agreed to in advance with the appropriate regulatory agencies. It is recognized that there may be practical limitations that affect the scope, methods, inputs, or outputs of the modeling:

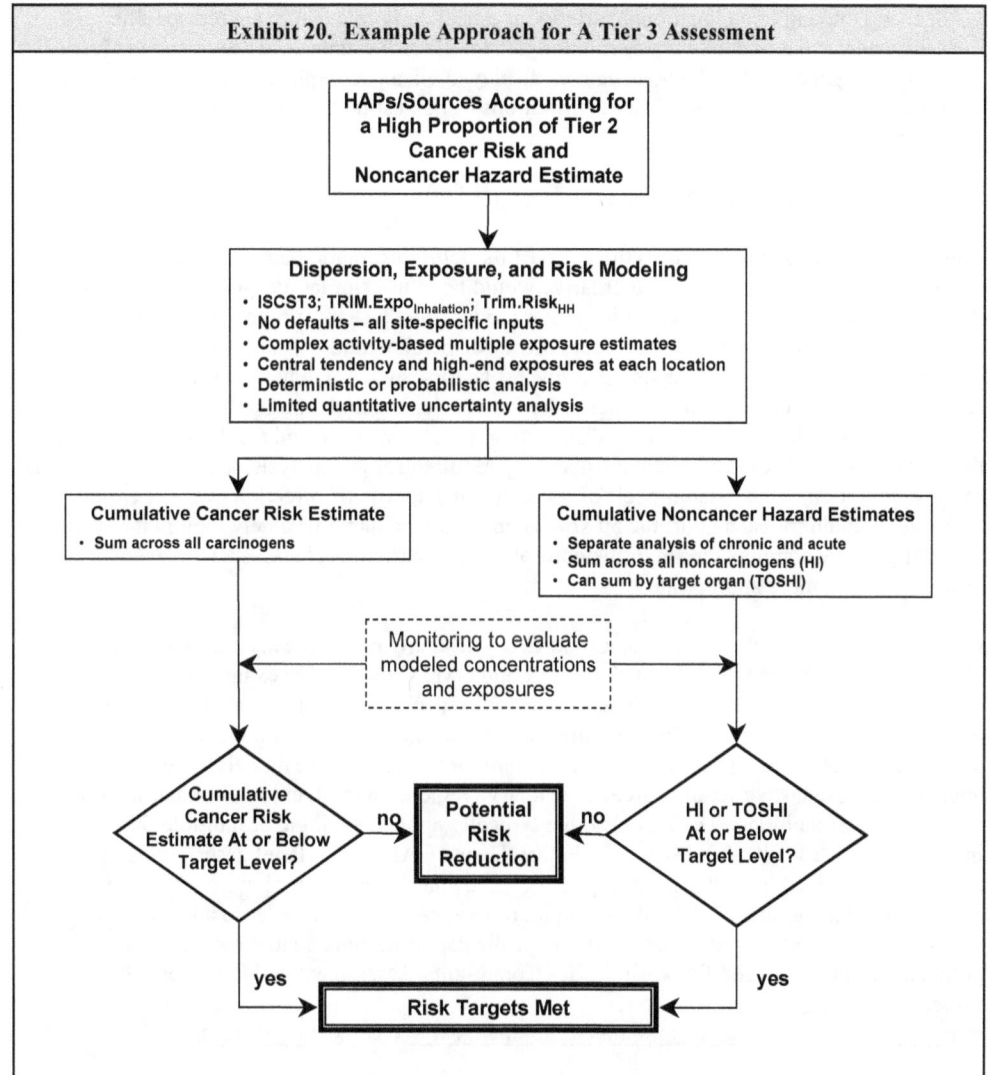

Exhibit 20. Example Approach for A Tier 3 Assessment

HAPs/Sources Accounting for a High Proportion of Tier 2 Cancer Risk and Noncancer Hazard Estimate

Dispersion, Exposure, and Risk Modeling
- ISCST3; TRIM.Expo$_{Inhalation}$; Trim.Risk$_{HH}$
- No defaults – all site-specific inputs
- Complex activity-based multiple exposure estimates
- Central tendency and high-end exposures at each location
- Deterministic or probabilistic analysis
- Limited quantitative uncertainty analysis

Cumulative Cancer Risk Estimate
- Sum across all carcinogens

Cumulative Noncancer Hazard Estimates
- Separate analysis of chronic and acute
- Sum across all noncarcinogens (HI)
- Can sum by target organ (TOSHI)

Monitoring to evaluate modeled concentrations and exposures

Cumulative Cancer Risk Estimate At or Below Target Level?

no → **Potential Risk Reduction** ← no

HI or TOSHI At or Below Target Level?

yes

yes

Risk Targets Met

- Because or computing and other practical constraints, it is considered unlikely that a distribution of ambient air concentration estimates will be developed for each location of interest.

- Quantitative information sufficient to support probabilistic modeling of dose-response parameters seldom exists.

- With regard to individual risk, presentation of distributions that reflect variation in both location as well as population characteristics can mask information that is important to decision making (e.g., locational variation is particularly relevant to source-specific assessment). Consequently, presentation of specific risk estimates across locations (e.g., central tendency and high end estimates displayed via a geographical information system) separate from presentation of any distributions of risk estimates for each location are preferred.

8.2   Fate and Transport Modeling

This example Tier 3 approach uses the ISCST3 model for estimating air concentrations for both chronic and acute exposures. An alternative would be to use similar air quality model, such as AERMOD, when it becomes available. Other models may be appropriate for a specific facility/source. The extent to which a specific air dispersion model is suitable for the evaluation of air toxic source impacts depends upon several factors, such as the nature of the pollutants (e.g., gaseous, particulate, reactive, inert), the meteorological and topographic complexities of the area, the complexity of the source distribution, the spatial scale and resolution required for the analysis, and the level of detail and accuracy required for the analysis. For example, because of the assumption in Gaussian models of a steady wind speed and direction over the entire modeling domain for each hour, the 50 km distance may be inappropriately long in many areas, especially where complex terrain or meteorology is present. In such cases a non-steady state model would be more appropriate.

Finer scale models, such as CAL3QHC and CALINE4, are most typically applied to exposure studies from mobile sources. The UAM-TOX and CMAQ models are examples of models which can simulate photochemically active air toxic species, including secondary formation of pollutants like formaldehyde. Because the complex secondary formation processes are nonlinear and can occur at locations far distant from the emission source, these models are designed to be applied to an exhaustive set of sources over a large region, rather than to individual facilities or small groups of facilities. The models more typically applied to single or multiple facilities include SCREEN3, ISCST3, ISCLT3, AERMOD, and CALPUFF. Brief descriptions of these models are provided in Volume 1 (Chapter 9). Some modeling studies have combined the application of a regional model with a neighborhood-scale model in order to address secondary and background concentration contributions, while capturing finer spatial resolution for primary pollutant predictions (see EPA's Air Toxics Community Assessment and Risk Reduction Projects Database
http://yosemite.epa.gov/oar/CommunityAssessment.nsf/Welcome?OpenForm).

Where the assessment could support a regulatory decision, the use of an alternative model commonly is agreed to in advance with the regulatory agency decision-maker. Alternative models that conform to EPA's air quality modeling guidance[17] are more likely to be acceptable to those decision-makers.

8.2.1   Model Inputs

Input data for ISCST3 falls into the same six categories used by HEM-3: (1) source location; (2) emissions data; (3) physical release parameters; (4) pollutant specific data (reactivity and dose-response data); (5) meteorological data; and (6) population data. Additional details are provided in the ISC model user's guide.[9]

- **Source location**. ISCST3 requires coordinate data (latitude and longitude) for each emission source in the analysis. Multiple sources can be modeled in the same ISCST3 run.

- **Emissions data**. The analysis generally benefits from the use of the highest-quality, site-specific emissions data available, including chemical speciation. With ISCST3, users have many more options for characterizing emissions. For example, users have the option to specify variable emission rate factors for sources whose emissions vary as a function of time (e.g., month, season, hour-of-day). In addition, settling velocity categories, mass fractions, and reflection coefficients may be specified for sources of large particulates that experience settling and removal during dispersion. *The emissions profile(s) used for Tier 3 modeling commonly reflect the expected pattern(s) of emissions over a reasonable period of time (e.g., several years)*. Note that these profiles may differ for different sources within a single facility.

- **Physical release parameters**. Tier 3 assessments commonly use facility/source-specific values for all stack/vent parameter data, as appropriate for the type of source. These values commonly reflect the expected patterns of emissions used to develop the emissions profiles for modeling.

- **Atmospheric reactivities of chemicals**. Tier 3 assessments often are focused on a relatively small number of HAPs and sources that collectively account for a large proportion of the risk and/or hazard estimated in lower-tier analyses. Therefore, it may be worthwhile to consider reactivity and deposition in Tier 3 assessments. However, because the highest inhalation exposures are generally near to sources, these processes may not substantially influence exposures unless they are very rapid. Exhibit 21 below provides a list of HAPs with atmospheric half-lives of one hour or less, for which considering reactivity may be useful. Volume I of this reference manual (Chapter 17) provides more data sources for half-life data. Atmospheric degradation in ISTST3 is limited to a first order approximation. ISCST3 can also consider wet and dry deposition, but requires source- and HAP-specific information (e.g., on particle size and mass and solubility). *As noted earlier, for HAPs with a significant potential to form reaction products in the atmosphere, either a different model should be used, or the analysis should note the added uncertainties in not modeling these reaction products.*

- **Meteorology**. Tier 3 analyses commonly use the most recent consecutive 5 years of facility/source-specific data from the nearest representative meteorological station. "Representative" generally means that the sources being modeled and the weather station are located in the same general environment with respect to *significant terrain* (e.g., valley vs. plateau), *significant geographic features* (e.g., proximity to a large body of water), and *prevailing winds* (i.e., similar wind rose/direction of dominant wind). Documentation demonstrating the representativeness of the meteorological data is encouraged. Sources of meteorological data are provided in Volume 1, Appendix G.

Exhibit 21. Volatile HAPs with Atmospheric Half-lives of Less than One Hour	
CAS Number	**Chemical Name**
107-02-8	Acrolein
107-05-1	Allyl chloride
106-99-0	Butadiene
126-99-8	Chloroprene
1319-77-3	Cresols/Cresylic acid (isomers and mixture)
68-12-2	Dimethyl formamide
140-88-5	Ethyl acrylate
111-76-2	Ethylene glycol butyl ether
80-62-6	Methyl methacrylate
90-12-0	Methylnaphthalenes
91-20-3	Naphthalene
108-95-2	Phenol
123-38-6	Propionaldehyde
100-42-5	Styrene
121-44-8	Triethylamine
108-05-4	Vinyl acetate
1330-20-7	Xylenes

- **Receptor Locations**. In Tier 3 these are set with consideration of the demographic unit resolution for the TRIM.Expo$_{Inhalation}$ exposure estimates. These may be specified as a list of receptor locations (e.g., census tract internal points). Alternatively, they may be specified as arrays of receptors (e.g., radially or via grid), and TRIM.Expo$_{Inhalation}$ has the capability to associate these air concentration locations with corresponding demographic units.

8.2.2 Model Runs

Using one complete year of meteorological data, ISCST3 calculates an ambient concentration for each receptor location (district) specified in a run, for each hour of a one-year run. The ISCST model can be run for single or multiple years. To obtain results for multiple years of meteorological data, separate ISCST3 model runs commonly are completed for each year. Based on these results, ISCST3 can provide average ambient concentration for each modeled location for the time periods of interest (e.g., 1 hour to 5 years).

These concentrations are used to prepare inventories of chemical concentrations at each specified exposure location at selected time intervals (e.g., days, hours). The specific ISCST3 model outputs used to prepare inventories of chemical concentrations may depend on how the analysis is structured. As a general guideline:

- For cancer effects, calculate an annual average concentration using the one-hour concentrations for each of the five years;

- For chronic noncancer effects, calculate an annual average concentration using the highest rolling annual average of one-hour concentrations for the five years; and

- For acute noncancer effects, identify the highest one-hour concentration in all five years.

8.3 Exposure Assessment

In this example Tier 3 approach, the exposure model (TRIM.Expo$_{Inhalation}$) is used to combine the air concentration estimates with human activity patterns to develop more precise, population- and site-specific exposure estimates. The user specifies the geographic area to be modeled, the number of individuals to be simulated to represent the study area population, and the demographic unit of resolution for the outputs. TRIM.Expo$_{Inhalation}$ generates a personal profile for each simulated person that specifies various parameter variables required by the model. The model next uses diary-derived time/activity data matched to each personal profile to generate an exposure event sequence (also referred to as an "activity pattern" or "composite diary") for the modeled individual that spans a specified time period, such as one year. Each event in the sequence specifies a start time, an exposure duration, a geographic location, a microenvironment, and an activity. Probabilistic algorithms are used to estimate the pollutant concentration associated with each exposure event. The estimated pollutant concentrations account for the effects of ambient (outdoor) pollutant concentration, penetration factors, air exchange rates, decay/deposition rates, and proximity to emission sources, depending on the microenvironment, available data, and the estimation method selected by the user. The model produces a distribution of exposure estimates for the individuals modeled for each demographic unit (e.g., census tract or block).

Major aspects of the set up and execution of the TRIM.Expo$_{Inhalation}$ model are described in separate subsections below. The TRIM.Expo$_{Inhalation}$ User's Document provides additional details.[10] Where the assessment could support a regulatory decision, advance discussions with the regulatory agency risk manager are encouraged to confirm the approach to be employed for the exposure assessment.

8.3.1   Characterization of the Study Area

The study area will have been considered in the design of the air quality modeling step, including the resolution desired for the demographic unit for which exposure estimates are to be generated. In the exposure modeling step, the user will confirm the extent of the study area for which air quality estimates are available, and the demographic unit scale at which exposure estimates are to be generated.

If the air quality modeling was designed to yield ambient concentration estimates associated with the demographic units of interest for the exposure assessment (e.g., air concentration estimates at census tract/block centroids), TRIM.Expo will use those locations directly for developing population specific exposure estimates. Otherwise, TRIM.Expo will assign the air quality estimates to demographic units based on the distance between the location of each air quality estimate and the geographic center of each demographic unit.

In TRIM.Expo$_{Inhalation}$, the geographic units for the demographic data are called **sectors**. The model uses the demographic data to create personal profiles at the sector level. For each sector the model requires demographic information representing the distribution of age, gender, race, and work status within the study population. The initial release of TRIM.Expo$_{Inhalation}$ has input files that already contain this demographic and location data for all Census tracts in the 50 United States based on the 2000 Census. This database enables the user to model any study area in the country without having to make any changes to these input files. Finer scales, such as Census block groups and blocks may be used with correspondingly suitable population data files. If fewer (thus larger) sectors were desired, the existing population data files could be aggregated to larger regions.

The spatial units for the ambient air quality estimates are called **districts**. These are the areas associated with each air quality estimate. As mentioned earlier, the air quality modeling may be designed so that the districts are the same as the sectors. For example, the air quality modeling may produce estimates for census tract centroids and census tracts may be the sectors to be used in the exposure modeling. If, however, the air quality estimate locations are not synonymous with the sector centers, TRIM.Expo will assign the air quality districts to sectors based on distance between their geographic centers (or other user-specified point). Each sector is assigned the air quality district for which this distance is shortest, yet within a user-specified maximum distance. Sectors falling outside that distance are not modeled.

Another spatial unit in TRIM.Expo is the temperature zone, which may be used by the model to assign activity diaries to personal profiles (e.g., summer activities in summer weather) for the modeled individuals (see Section 8.3.2), and to specify parameters pertaining to microenvironments (e.g., aspects of climate control systems). The demographic sectors are assigned to temperature zones in similar fashion to the air quality district assignments (e.g., by closest proximity of user-specified representative locations

In summary, the exposure assessment **study area** is composed of the demographic sectors for which air quality estimates (and temperature values) can been assigned.

8.3.2 Generation of Simulated Individuals

TRIM.Expo_Inhalation probabilistically generates a user-specified number of simulated persons to represent the population in the study area. Each simulated person is represented by a "personal profile." TRIM.Expo_Inhalation generates the simulated person or profile by probabilistically selecting values for a set of profile variables (Exhibit 22). The profile variables include:

- Demographic variables, which are generated based on the Census data;
- Residential variables, which are generated based on sets of distribution data;
- Daily varying variables, which are generated based on distribution data that change daily during the simulation period; and
- Physiological variables, which are generated based on age group-specific distribution data.

Exhibit 22. Profile Variables in TRIM.Expo_Inhalation		
Variable Type	**Profile Variables**	**Description**
Demographic variables	Gender	Male or Female
	Race	White, Black, Native American, Asian, Other
	Age	Age of the simulated person (years)
	Home sector	Sector in which the simulated person lives
	Work sector	Sector in which the simulated person works
	Employment status	Indicates employment outside home
Residential variables	Gas stove	Indicates presence of gas stove
	Gas pilot	Indicates presence of gas pilot light
	Air conditioner	Indicates presence of air conditioning at home
	Car air conditioner	Indicates presence of air conditioning in the car
Daily varying variables	Window position	Daily window position (open or closed) during the simulation period
	Daily average car speed	Daily average car speed during the simulation period

TRIM.Expo_Inhalation first selects and calculates demographic, residential, and physiological variables (except for daily values) for all the specified number of simulated individuals, and then determines exposures for each simulated person. The exposure assessor should be sure to simulate enough individuals to produce a stable exposure distribution (i.e., one that is not sensitive to which specific types of individuals the model randomly selects). The TRIM.Expo_Inhalation User's Document provides additional details.[10]

8.3.3 Construction of A Sequence of Activity Events

TRIM.Expo$_{Inhalation}$ probabilistically creates a composite diary for each of the simulated persons by selecting a 24-hour diary record (a **diary day**) from an activity database for each day of the simulation period. This diary is created based on the activity data in the Consolidated Human Activities Database (CHAD).[18] CHAD data have been supplied with TRIM.Expo$_{Inhalation}$ for this purpose. A composite diary is a sequence of events that simulate the movement of a modeled person through geographical locations and microenvironments during the simulation period. Each event is defined by geographic location, start time, duration, microenvironment visited, and an activity performed. For the activity database, TRIM.Expo$_{Inhalation}$ currently provides a personal information file and an events file to summarize the CHAD data. These composite diaries are then used in exposure concentration calculations.

8.3.4 Calculation of Concentrations in Microenvironments

TRIM.Expo$_{Inhalation}$ calculates ambient air concentrations in the various microenvironments visited by the simulated person by using the ambient air data for the relevant sectors and the user-specified method and parameters that are specific to each microenvironment.

TRIM.Expo$_{Inhalation}$ defines microenvironments by grouping the more than100 location codes defined in the activity (CHAD) database into a smaller set of user-defined microenvironments amenable to modeling. The user has control over how many microenvironments will be modeled, how they are defined, and what CHAD (or other activity database) locations should be grouped into each of microenvironment. The TRIM.Expo$_{Inhalation}$ User's Document[10] lists the 115 CHAD location codes included in TRIM.Expo$_{Inhalation}$ and the microenvironment to which each currently is assigned.

TRIM.Expo$_{Inhalation}$ calculates concentrations of the subject air pollutant in all the microenvironments at each time step (e.g., hour) of the simulation period for each of the simulated individuals, based on the user-provided hourly ambient air quality data specific to the geographic locations visited by the individual. TRIM.Expo$_{Inhalation}$ provides two methods for calculating microenvironmental concentrations: a **mass balance** method and a **factors** method (see Volume 1, Chapter 11). The user is required to specify a calculation method for each of the microenvironments; some microenvironments can use one method while the rest use the other, without restrictions. The parameters, algorithms, and probabilistic elements used in each of the methods are explained in the TRIM.Expo$_{Inhalation}$ User's Document.[10]

8.3.5 Estimating Exposure

TRIM.Expo$_{Inhalation}$ calculates exposure as a time series of exposure concentrations that a modeled individual experiences during the simulation period. TRIM.Expo$_{Inhalation}$ determines the exposure based on the air concentration in and minutes spent in each of a sequence of microenvironments visited according to the composite diary. The exposure concentration at any clock hour during the simulation period is determined using the following equation:

$$C_i = \frac{\sum\limits_{j=1}^{N} \left(C_{hourly(j)} \, t_{(j)} \right)}{T}$$

where:

C_i = Hourly exposure concentration at clock hour I of the simulation period ($\mu g/m^3$ or ppm)
N = Number of events (i.e., microenvironments visited) in clock hour I of the simulation period.
$C_{hourly(j)}$ = Hourly concentration in microenvironment j ($\mu g/m^3$ or ppm)
$t_{(j)}$ = Time spent in microenvironment j (minutes)
T = 60 minutes

From the hourly exposures, TRIM.Expo$_{Inhalation}$ calculates time series of 1-hour, 8-hour and daily average exposure concentrations that a simulated individual would experience during the simulation period. TRIM.Expo$_{Inhalation}$ then statistically summarizes and tabulates the hourly, 8-hour, and daily exposures. TRIM.Expo$_{Inhalation}$ also calculates and tabulates an annual average exposure concentration per individual. *The average concentration over the entire simulation period is used as an input to chronic risk calculations.*

8.3.6 General Considerations

Regardless of whether a deterministic or probabilistic approach is used, this example Tier 3 exposure assessment provides a range of exposure estimates, as evidenced by the exposures predicted for the multiple simulated individuals in each exposure district. This information facilitates presentation of central-tendency and high-end exposure estimates.

- Probabilistic analyses include frequency distributions of exposures that include central-tendency and high-end exposures for all locations and population sub-groups.

- Additionally, different estimates of ambient air concentrations (e.g., annual average vs. an alternate estimate) might be considered.

8.3.7 Monitoring Data

It may be helpful to conduct a monitoring program to evaluate or further characterize exposure concentrations, such as through the use of air measurements at key locations (e.g., to establish indoor or outdoor air concentrations) or exposure concentrations associated with specific activity patterns (e.g., through the use of personal monitors). Volume 1 (Chapter 10) provides an overview of available approaches for assessing air quality via monitoring.

8.4 Risk Characterization

For Tier 3, most risk estimates will be calculated automatically by TRIM.Risk$_{HH}$, once the modeling scenarios are set up and run. The initial release of TRIM.Risk$_{HH}$ will calculate several inhalation cancer risk and non-cancer hazard metrics at the individual and population level.

This example Tier 3 risk characterization considers background concentrations as appropriate (e.g., where the contribution of facility/source to risk is small compared to that of other sources. *Note that consideration of background may or may not be appropriate pursuant to the specific legal and regulatory authorities under which the risk assessment is being conducted.*

8.4.1 Reporting Results

The initial release of TRIM.Risk$_{HH}$ will provide a visualization tool for presenting analysis results in various automated formats. The risk assessor may want to utilize this tool to assist with development of the risk characterization summary, which generally will include the following information:

- Documentation of input parameters, outputs, and risk characterization, with special emphasis on the range of risk or hazard estimates;

- A simple presentation describing the assessment's purpose and the outcome relative to the purpose (e.g., purpose: demonstrate that risk is below target levels; outcome: low risk not demonstrated); and

- Documentation of all key assumptions or other inputs used for the assessment, such as emissions characteristics or choice of a nearby meteorological station. In particular, the assessor is encouraged to document the rationale used to define personal profiles and how they are modeled and analyzed.

8.4.2 Assessment and Presentation of Uncertainty

Risk managers need to understand the strengths and the limitations of the Tier 3 assessment. A critical part of the risk characterization process, therefore, is an evaluation of the assumptions, limitations, and uncertainties inherent in the Tier 3 risk assessment in order to place the risk estimates in proper perspective.[3] Tier 3 risk assessments commonly include semi-quantitative sensitivity analyses and quantitative uncertainty analysis (described in Section 5.4 above). The general quantitative approach to propagating or tracking uncertainty through probabilistic modeling is described in Volume 1 (Chapter 31) of this reference library.

Chapter IV: Multipathway Risk Assessment

This section constitutes a snapshot of EPA's current thinking and approach to the adaptation of the evolving methods of multipathway risk assessment to the context of Federal and state control of air toxics. While inhalation risk assessment has been increasingly used in regulatory contexts over the last several years, multipathway risk assessment tools are less well developed and field tested in a regulatory context. This section should be considered a living document for review and input. By publishing this portion of Volume 2 in its current state of development, EPA is soliciting the involvement of persons with experience in this field to help improve these assessment methods for use in a regulatory context. EPA anticipates revisions to this draft section of Volume 2 on the basis of this input.

1.0   Introduction and Overview

This section describes an example approach for performing site-specific multipathway risk assessments for air toxics. These include an assessment of human health risks via ingestion (and perhaps other pathways/routes) as well as ecological risk assessment. Both types of assessments consist of a tiered approach. A risk assessor may decide to complete only the lowest-tier analysis that fits the purpose of the assessment (e.g., to determine that a facility's cumulative risk is lower than a risk manager's level of concern). Conversely, an assessor may choose not to complete a lower-tier analysis before completing a higher-tier analysis (e.g., the risk assessor could go directly to Tier 3).

In this example approach, a multipathway risk assessment is conducted only if PB-HAP compounds are present in facility/source emissions (see Exhibit 23), and acute exposures are not assessed for multipathway analyses (although they may be for inhalation analyses). The focus is on exposure and risk/hazard once the PB-HAPs are deposited onto soils, into surface waters, etc. Other situations may warrant a multipathway analysis, such as when there are significant emissions of pollutants whose primary risk is through non-inhalation exposure pathways, or when state or local regulations require such analyses.

The discussion in this chapter is divided into the following sections:

*   Section 2 provides an overview of the multipathway risk assessment and the tiered approaches to the human health and ecological analyses;

*   Section 3 provides information for preparing the emissions inventory for the multipathway assessment;

*   Section 4 describes an example approach for the multipathway human health risk assessment; and

*   Section 5 provides an example approach for the ecological risk assessment.

Exhibit 23. HAPs of Concern for Persistence and Bioaccumulation (PB-HAP Compounds)			
PB-HAP Compound	**Pollution Prevention Priority PBTs**	**Great Waters Pollutants of Concern**	**TRI PBT Chemicals**
Cadmium compounds		X	
Chlordane	X	X	X
Chlorinated dibenzodioxins and furans	X[a]	X	X[b]
DDE	X	X	
Heptachlor			X
Hexachlorobenzene	X	X	X
Hexachlorocyclohexane (all isomers)		X	
Lead compounds	X[c]	X	X
Mercury compounds	X	X	X
Methoxychlor			X
Polychlorinated biphenyls	X	X	X
Polycyclic organic matter	X[d]	X	X[e]
Toxaphene	X	X	X
Trifluralin			X
[a] "Dioxins and furans" [b] "Dioxin and dioxin-like compounds" [c] Alkyl lead [d] Benzo[a]pyrene [e] "Polycyclic aromatic compounds" and benzo[g,h,i]perylene			

2.0   Tiered Approach and Models Used

The example multipathway risk assessment approach currently includes only two tiers of analysis (Tier 2 and Tier 3, see Exhibit 24). EPA is investigating whether it is possible to develop a Tier 1 methodology for multipathway analyses. *Therefore, at this point in time, this example multipathway analysis starts with a Tier 2-level approach*.

- Tier 1 is conceptualized as a simple look-up table or graph that identifies threshold emissions rates (tons/year) for each PB-HAP compound below which multipathway risks are not of concern. The objective would be to allow a facility/source that emits small amounts of PB-HAP compounds to demonstrate that risk targets are met without the need for facility-specific modeling (as, for example, in a Tier 2 analysis). However, at this point in time, EPA does not have sufficient experience with multipathway air toxics risk assessments to develop a Tier 1 approach.

- The example Tier 2 approach uses ISCST3 and the **Methodology for Assessing Health Risks Associated with Multiple Pathways of Exposure to Combuster (MPE)** to estimate concentrations in air, water, soil, and biota. For the human health assessment, MPE is used to estimate cancer risk and noncancer hazard to a hypothetical receptor characterized using a conservative "subsistence farmer" scenario. For the ecological risk assessment, concentrations in air, water, and biota are compared to media-specific ecological toxicity reference values (i.e., a generic set of ecological receptors is assumed).

- The example Tier 3 approach uses the Total Risk Integrated Methodology (TRIM) in a deterministic or stochastic mode. TRIM **Fate, Transport, and Ecological Exposure (TRIM.FaTE)** is used to estimate concentrations in air, water, soil, and biota. For the human health risk assessment, $TRIM.Expo_{Ingestion}$ (coupled with a farm food chain model) is used to derive estimates of ingestion exposure for a set of scenarios (e.g., resident, farmer, fisher), and $TRIM.Risk_{HH}$ is used to derive cancer risk and chronic noncancer hazard estimates. For the ecological risk assessment, $TRIM.Risk_{Eco}$ is used to derive estimates of ecological risk by comparing estimated concentrations of PB-HAPs in abiotic media, dietary intakes, and body burdens to corresponding ecological toxicity reference values.

Although it is possible to evaluate acute exposures for the ingestion pathway, EPA does not generally recommend such assessments for substances released to the air because it is very unlikely that acute ingestion threats could exist under typical release conditions and in the absence of serious chronic ingestion risks. However, each assessment should consider the available evidence in making this judgment. The risk assessor is encouraged to state the reasons why an acute analysis for non-inhalation pathways was not performed.

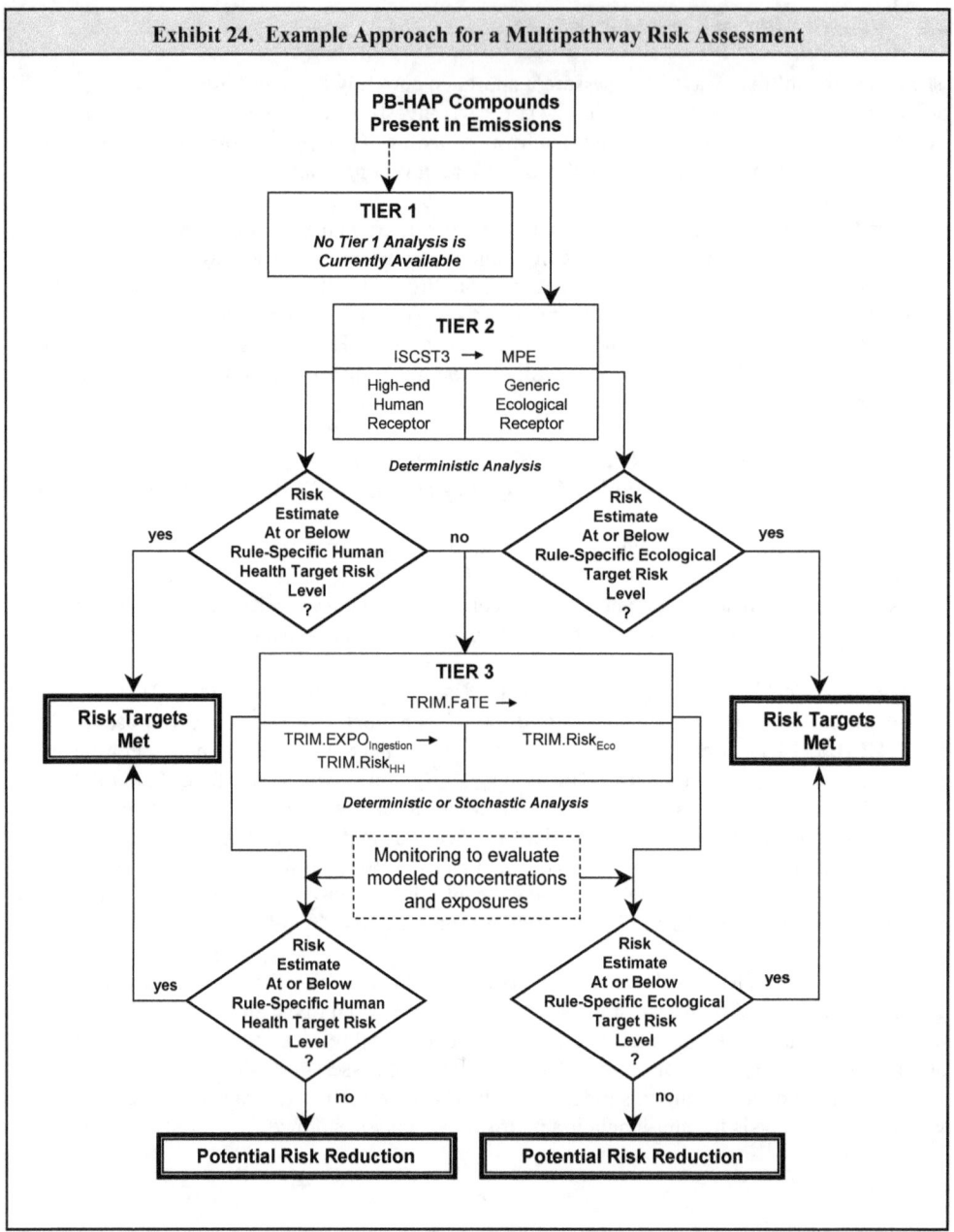

Exhibit 24. Example Approach for a Multipathway Risk Assessment

PB-HAP Compounds Present in Emissions

TIER 1
No Tier 1 Analysis is Currently Available

TIER 2
ISCST3 → MPE

| High-end Human Receptor | Generic Ecological Receptor |

Deterministic Analysis

Risk Estimate At or Below Rule-Specific Human Health Target Risk Level ?

Risk Estimate At or Below Rule-Specific Ecological Target Risk Level ?

yes — no — yes

Risk Targets Met

Risk Targets Met

TIER 3
TRIM.FaTE →

| TRIM.EXPO_Ingestion → TRIM.Risk_HH | TRIM.Risk_Eco |

Deterministic or Stochastic Analysis

Monitoring to evaluate modeled concentrations and exposures

Risk Estimate At or Below Rule-Specific Human Health Target Risk Level ?

Risk Estimate At or Below Rule-Specific Ecological Target Risk Level ?

yes — yes

no — no

Potential Risk Reduction

Potential Risk Reduction

Overview of Models Cited in this Example Approach

MPE. MPE, formerly referred to as the Indirect Exposure Methodology (IEM), primarily consists of a set of multimedia fate and exposure algorithms developed by EPA's Office of Research and Development.[19][e] The *Human Health Risk Assessment Protocol for Hazardous Waste Combustion Facilities*, under development by EPA's Office of Solid Waste, is an implementation of this approach.[20] The MPE approach includes procedures for estimating human exposures and health risks resulting from the transfer of emitted pollutants from air to soil and surface water bodies and the subsequent uptake by vegetation, animals, and humans. The methodology specifically addresses exposures via inhalation; ingestion of food, water, and soil; and dermal contact. The MPE methodology was designed to predict long-term, steady-state impacts from continuous sources, rather than short-term, time-series estimates. It consists of a "one-way process" through a series of linked models and algorithms, beginning with the modeling of the transport of pollutant emissions in air and the subsequent deposition to soil and surface water and culminating in the uptake of the emitted pollutant(s) into biota.

TRIM.FaTE. TRIM.FaTE is a spatially explicit, compartmental mass balance model that describes the movement and transformation of pollutants over time, through a user-defined, bounded system that includes both biotic and abiotic components (compartments).[21] The TRIM.FaTE module predicts pollutant concentrations in multiple environmental media and in biota and pollutant intakes for biota, all of which provide both temporal and spatial exposure estimates for ecological receptors (i.e., plants and animals). The output concentrations from TRIM.FaTE can also be used as inputs to a human ingestion exposure model, such as TRIM.Expo$_{Ingestion}$, to estimate human exposures.

TRIM.Expo$_{Ingestion}$. TRIM.Expo$_{Ingestion}$ calculates the ingestion exposure to human receptor groups from media and food concentrations estimated using TRIM.FaTE output data (or other pollutant concentration data) for media and biota. A farm food chain module provides livestock and produce contaminant estimates from air and soil concentrations and air deposition estimates provided by TRIM.FaTE or from an external file.

TRIM.Risk$_{HH}$. In TRIM.Risk$_{HH}$, estimates of human exposures are characterized with regard to potential risk using the corresponding exposure- or dose-response relationships. The output from TRIM.Risk$_{HH}$ includes documentation of the input data, assumptions in the analysis, and the results of risk calculations and exposure analysis.

TRIM.Risk$_{Eco}$. In TRIM.Risk$_{Eco}$, estimates of ecological exposures are characterized with regard to potential risk using the corresponding exposure- or dose-response relationships. The output from TRIM.Risk$_{Eco}$ includes documentation of the input data, assumptions in the analysis, and the results of risk calculations and exposure analysis.

Note: As of the publication of this document, Trim.Expo and Trim.Risk are not yet available. Documentation for the current versions of all of these models may be found on the Fate, Exposure, and Risk Analysis (FERA) page (http://www.epa.gov/ttn/fera/).

---

[e] Note that the MPE model and many of its variations are conceptual models used to describe fate and transport, not "ready-to-run" computer models. Typically, users incorporate these conceptual models into spreadsheets or other computer frameworks to create a usable model.

3.0 Developing the Emissions Inventory for Multipathway Analyses

As with the inhalation analysis, developing the emissions inventory involves (1) quantifying emissions rates; (2) quantifying other emissions parameters important for the exposure assessment (e.g., temperature, stack height); (3) identifying the chemical species of the emitted HAPs (where applicable); and (4) identification of background concentrations of the HAPs being released (in some instances).

As with inhalation, the risk assessor is encouraged to use its highest quality, most detailed emissions data. Acceptable data, in order by preference, include (1) actual measured emissions from a recent high-activity, high-emission year, (2) measured emissions extrapolated to a high-activity, high-emission year, (3) facility/source-specific engineering estimates of a high-activity, high emission year (with documentation); and (4) permitted emissions, (e.g., the maximum allowed under MACT, or under a permit) with documentation that the permitted limits are not exceeded.

In general, if lower quality data (i.e., of types (3) and (4) above) are used, these data commonly represent high-end estimates of emissions in order to ensure the assessment will produce health-protective results. The inventories commonly include the following release parameters for each HAP released from each source: volume, schedule, emission factors, and applicable time periods. Emissions commonly represent conditions typical of a high-activity, high-emission year.

In this example approach, the risk assessments consider emissions controls in use at the facility/source. At the least, the default assumption is that the facility's sources are in compliance with the appropriate MACT standards and permit requirements, although it may be reasonable to modify this assumption if additional emissions controls are in place. The documentation for the assessment includes the specific emissions inventories used.

4.0 Multipathway Human Health Risk Assessment

4.1 Introduction and Overview

The multipathway human health risk assessment is divided into two tiers, beginning with Tier 2:

- This example Tier 2 analysis uses the ISCST3 model and the MPE methodology[19] to estimate risk/hazard at the most impacted location via several ingestion pathways. The *Human Health Risk Assessment Protocol for Hazardous Waste Combustion Facilities* (HHRAP),[20] under development by EPA's Office of Solid Waste, is an implementation of the MPE approach. Modeled deposition rates (from ISCST3) are converted by MPE into chemical concentrations in media and biota. MPE then uses exposure scenarios (e.g., resident, farmer, fisher) to estimate exposure. Monitoring is not included explicitly in this example Tier 2 analysis because it is intended as a health-protective analysis.

- This example Tier 3 analysis uses TRIM.FaTE, TRIM.Expo$_{Ingestion}$, and TRIM.Risk$_{HH}$, to estimate ingestion risk/hazard (see Exhibit 25). The initial version of TRIM.Expo$_{Ingestion}$ uses a scenario-based approach to model exposure. In TRIM Risk$_{HH}$, estimates of human ingestion exposures (intake rates) are characterized with regard to potential risk using the corresponding dose-response relationships. As noted in Exhibit 25, measured concentrations from monitoring programs can be used as inputs to TRIM.Expo and thus can be incorporated into the Tier 3 analysis.

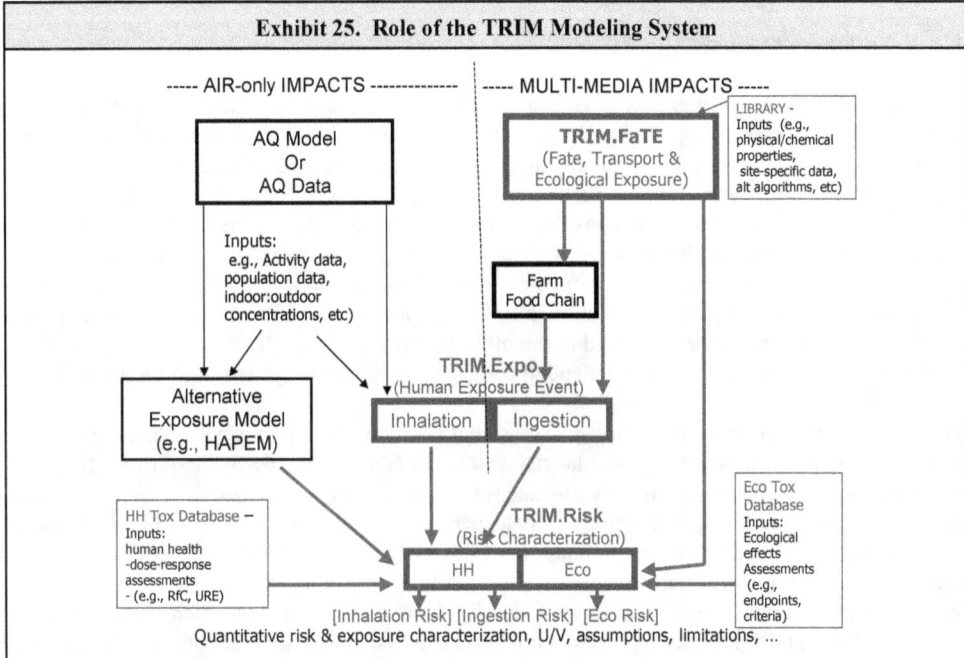

Exhibit 25. Role of the TRIM Modeling System

The Total Risk Integrated Methodology (TRIM) modeling system can be used to assess human inhalation, human ingestion, and ecological risks. TRIM.FaTE accounts for movement of a chemical through a comprehensive system of discrete compartments (e.g., media and biota) that represent possible locations of the chemical in the physical and biological environments of the modeled ecosystem and provides an inventory, over time, of a chemical throughout the entire system. In addition to providing exposure estimates relevant to ecological risk assessment, TRIM.FaTE generates media concentrations relevant to human ingestion exposures that can be used as input to the ingestion component of the Exposure-Event module, TRIM.Expo. Measured concentrations also can be used as inputs to TRIM.Expo. In the inhalation component of TRIM.Expo, human exposures are evaluated by tracking randomly selected individuals that represent an area's population and their inhalation and ingestion through time and space. Trim.Expo$_{Inhalation}$ can accept ambient air concentration estimates from an external air quality model or monitoring data. In the Risk Characterization module, TRIM.Risk estimates of human exposures or doses are characterized with regard to potential risk using the corresponding exposure- or dose-response relationships. The TRIM.Risk module is also designed to characterize ecological risks from multimedia exposures. The output from TRIM.Risk is intended to include documentation of the input data, assumptions in the analysis, and metrics of uncertainty/variability, as well as the results of risk calculations and exposure analysis.

Both of these tiers of analysis require significant effort and input data, with many facility/source-specific considerations and parameter values. This document provides a brief description of the general modeling approaches and required input data, but a detailed discussion is beyond the scope of this document. Such information can be found in user's guides and related documentation on EPA's Fate, Exposure, and Risk Analysis (FERA) web page (http://www.epa.gov/ttn/fera/). Where the risk assessment involves a regulatory decision, the risk assessor is encouraged to discuss the scope, modeling/monitoring approach, and input data with the appropriate regulatory authorities prior to conducting the analysis.

4.2 Pathways Evaluated

The example multipathway exposure assessment in this technical resource document focuses on two general categories of ingestion pathways. Incidental ingestion pathways consider exposures that may occur from ingestion of soils or surface water while engaged in other activities (e.g., ingestion of soil while gardening or playing outside; ingestion of surface water while swimming). Food chain pathways consider exposures that may occur if PB-HAP compounds accumulate in the food and water people consume. Exhibit 26 provides some examples of which specific pathways could be evaluated for each PB-HAP compound in a multipathway human health assessment. As a general guideline, all of the pathways identified in the conceptual model for the assessment should be evaluated in the multipathway analysis. Note also that state and local agencies may have other recommended compounds and pathways required for analysis.

The focus of this example multipathway assessment is on ingestion pathways. Other exposure pathways may be important for particular risk assessments, including dermal exposures (i.e., direct contact with contaminated soils, surface waters, or surface water sediments during outside activities such as gardening or swimming); resuspension of dust (e.g., from wind blowing across contaminated soils) and subsequent inhalation of the dust particles; and ingestion of contaminated groundwater. However, EPA does not have sufficient experience with multipathway air toxics risk assessments to identify the circumstances for which exposures via these additional pathways may represent a potential concern. Each assessment may consider the available evidence in determining whether the risk assessment should include dermal exposure, resuspension of dust, and ingestion of groundwater. At a minimum, the risk characterization should state the reasons why an acute analysis for non-inhalation pathways was not performed.

- If facility/source-specific circumstances suggest that dermal pathways may be of concern, EPA's *Risk Assessment Guidance for Superfund (RAGS), Part E, Supplemental Guidance for Dermal Risk Assessment*,[22] includes a relatively straightforward methodology for dermal exposure and risk assessment, starting with soil concentrations. The Planning Tables in the document are simple to use and incorporate into the multipathway analysis.

- Relative to the direct inhalation pathway, inhalation of soil resulting from dust resuspension by wind erosion generally is not a significant pathway of concern for air toxics risk assessments. If facility/source-specific circumstances suggest that resuspension of dust may represent a potential concern, EPA's MPE methodology[19] discusses the methods for evaluating this pathway.

- If facility/source-specific circumstances suggest that groundwater may represent a potential concern (e.g., the presence of extremely shallow aquifers used for drinking water purposes, or a karst environment in which the local surface water significantly affects the quality of ground water used as a drinking water source) the TRIM.FaTE library includes a groundwater compartment that can and has been used to assess the groundwater pathway. EPA's *Human Health Risk Assessment Protocol for Hazardous Waste Combustion Facilities*[20] and *Draft Technical Background Document for Soil Screening Guidance*[23] discuss the methods for evaluating the groundwater pathway.

Exhibit 26. Examples of Specific Pathways to be Analyzed for Each PB-HAP Compound								
PB-HAP Compound	Soil ingestion	Dermal contact	Meat, milk, and eggs ingestion	Fish ingestion	Fruit and vegetable ingestion	Root vegetable ingestion	Water ingestion	Breast milk ingestion
Cadmium compounds	•	•			•	•	•	
Chlordane	•	•	•	•	•	•		•
Chlorinated dibenzodioxins and furans	•	•	•	•	•	•		•
DDE	•	•	•	•	•	•		
Heptachlor	•	•	•	•	•	•		•
Hexachlorobenzene	•	•	•	•	•	•		•
Hexachlorocyclohexane (all isomers)	•	•	•	•	•	•		•
Lead compounds[a]								
Mercury compounds	•		•	•	•	•	•	
Methoxychlor	•	•	•	•	•	•		
Polychlorinated biphenyls	•	•	•	•	•	•		•
Polycyclic organic matter	•	•	•	•	•	•		(b)
Toxaphene	•	•	•	•	•	•		•
Trifluralin	•	•	•	•	•	•		

[a]EPA suggests assessing effects of lead exposures with the Integrated Uptake/Biokinetic (IEUBKK) model, which predicts blood lead levels in children exposed to lead concentrations in air, water, soil, and house dust.
[b]Most POM compounds do not tend to accumulate to high levels in breast milk. Polybrominated biphenyl ethers (PBDEs), which fit within the POM group definition in the CAA and are therefore HAPs, are a significant exception. When assessing PBDEs, breast milk should be included.

4.3   Estimating Dietary Intake

The following generic equation is used calculate dietary chemical intake:[24] Volume 1 of this Reference Library (Chapter 20) provides a more detailed discussion of this equation.

$$I = \frac{EC \times CR}{BW} \times \frac{EF \times ED}{AT}$$

where

I =   Chemical intake rate, or the amount of pollutant ingested per unit time per unit body mass, expressed in units of mg/kg-day. For evaluating exposure to noncarcinogens, the intake is referred to as Average Daily Dose (*ADD*); for evaluating exposure to carcinogenic compounds, the intake is referred to as Lifetime Average Daily Dose (*LADD*)

Chemical-related variable:

EC =   Exposure concentration of the chemical in the medium of concern for the time period being analyzed, expressed in units of mg/kg for soil and food or mg/L for surface water or beverages (including milk)

Variables that describe the exposed population (also termed "intake variables"):

CR =   Consumption rate, the amount of contaminated medium consumed per unit of time or event (e.g., kg/day for food items and L/day for water)
EF =   Exposure frequency (number of days exposed per year)
ED =   Exposure duration (number of years exposed)
BW =   Average body weight of the receptor over the exposure period (kg)

Assessment-determined variable:

AT =   Averaging time, the period over which exposure is averaged (days). For carcinogens, the averaging time is 25,550 days, based on an assumed lifetime exposure of 70 years; for noncarcinogens, averaging time equals ED (years) multiplied by 365 days per year

- The **exposure concentration (EC)** for a chemical is calculated separately for each food item and environmental medium of concern. The value of these variables is determined primarily by modeling; however, monitoring data may be used to evaluate or further characterize exposure concentrations at key locations considered (as noted earlier, where the risk assessment involves a regulatory decision, the risk assessor is encouraged to discuss the scope and approach for collecting monitoring data with the appropriate regulatory authorities prior to conducting the analysis).

- **Consumption rate (CR)** is the amount of contaminated food or medium consumed per event or unit of time (e.g., amount of fish consumed per meal or per day). Note that consumption rate changes with age (e.g., children and adults eat different amounts of the same food items). The consumption rate is usually calculated by first estimating the amount of a given medium consumed per unit time and then multiplying by a fraction of the total dietary intake for this type of food or medium (e.g., 25 percent), representing the amount consumed from the study area. The specific fraction applied depends on the assumptions used for the analysis and can range from zero to 100 percent. For example, if the analysis assumed an individual consumed 6 grams of fish per day and obtained 25 percent of the fish from the contaminated area, CR would be $6 \times .25 = 1.5$ g/day.

- The specific **exposure frequency (EF)** specifies the number of days exposed each year, which generally ranges from a weekly or seasonal basis to 350 days/year (i.e., a person is resident in an area but spends two weeks each year in a different location, such as on vacation).

- **Exposure duration (ED)** is the length of time that a particular assessment represents (e.g., a lifetime or a particular residence time). This is specified for each analysis.

- The choice of **body weight (BW)** for use in the exposure assessment depends on the definition of the person at potential risk. Because children have lower body weights, typical ingestion exposures per unit of body weight, such as for soil, milk, and fruits, are substantially higher for children. If a lifetime exposure duration (or an exposure duration over the childhood and adult years) is being evaluated, it needs to be based on differing values for the different age groups. If less than a lifetime exposure estimate is being evaluated, it is important to include the children's age group in the specific scenarios or cohorts used (see Volume 1, Chapter 20).

- **Averaging time.** When evaluating exposure to noncarcinogenic toxicants, intake is averaged over the period of exposure. For carcinogens, intakes are traditionally calculated by prorating the total cumulative dose over a lifetime (i.e., chronic daily intakes, also called lifetime average daily intakes).

4.4   Ingestion Toxicity Assessment

As noted in Volume 1 of the reference library (Chapter 12), toxicity assessment is accomplished in two steps: hazard identification and dose-response assessment. Dose-response values (e.g., CPFs, RfDs) are used to estimate the potential for adverse impacts resulting from exposure to a given concentration of a PB-HAP. Identifying critical human health endpoints (cancer vs. noncancer) and target organs is critical for structuring the multipathway risk assessment, including determining what ingestion exposure pathways are of potential concern and how to sum the risks from exposure to multiple HAPs. Volume 1 describes this process in greater detail. For each PB-HAP included in a risk assessment, the risk assessor should identify the critical human health endpoints and target organs to ensure that cumulative risk across all HAPs is estimated in a manner consistent with risk assessment principles.

EPA/OAQPS has developed a set of recommended screening-level chronic human health dose-response values for many HAPs. This information is presented in Appendix C of Volume 1, which provides information on the type of hazard associated with each HAP (e.g., cancer, non-cancer) and the applicable dose-response values for each HAP (e.g., RfCs, IURs). The most up-to-date list of default screening level dose-response values recommended by EPA for the 188 HAPs is provided at http://www.epa.gov/ttn/atw/toxsource/summary.html. Other dose-response values that have undergone independent peer review may also be acceptable, but generally should be consistent with EPA risk assessment guidelines and agreed upon in advance for assessments that have regulatory implications. Descriptive information on the type of health hazards associated with each HAP (e.g., cancer, noncancer) may be found at http://www.epa.gov.ttn/atw/haptindex.html.

The **oral cancer slope factor (CSF)** is used to estimate ingestion cancer risk. It is derived in a similar way as the inhalation unit risk estimate (IUR) (see Volume I of this reference library). The CSF is defined as the upper-bound excess lifetime cancer risk estimated to result from continuous exposure to an agent. The true risk to humans, while not identifiable, is not likely to exceed the upper-bound estimate (the CSF). The CSF is presented as the risk of cancer per mg of intake of the substance per kg body weight per day (i.e., $[\text{mg/kg-day}]^{-1}$).

The oral **Reference Dose (RfD)** is used to estimate hazard. The RfD is expressed as a chronic dietary intake level (in units of mg/kg-day). The RfD is an estimate (with uncertainty spanning perhaps an order of magnitude) of a continuous exposure to the human population (including sensitive subgroups) that is likely to be without an appreciable risk of deleterious effects during a lifetime. In other words, exposures at or below the RfD will probably not cause adverse health effects, even to sensitive subgroups. This generally is used in EPA's health effects assessments for effects other than cancer.

Definition of Terms in Ingestion Dose-Response Values

Oral Cancer Slope Factor (CSF): An upper bound, approximating a 95% confidence limit, on the increased cancer risk from a lifetime exposure to an agent. This estimate is usually expressed in units of proportion (of a population) affected per mg/kg-day.

Reference Dose (RfD): An estimate (with uncertainty spanning perhaps an order of magnitude) of a daily oral exposure to the human population (including sensitive subgroups) that is likely to be without an appreciable risk of deleterious effects during a lifetime.

4.5   Risk Characterization for Ingestion Analysis

The process for characterizing cancer risks and noncancer hazards in this example multipathway analysis approach can be thought of as developing information to fill in a matrix similar to that shown in Exhibit 27 in addition to the discussion of assumptions, limitations, and uncertainties that are an essential part of risk characterization. A table like this would be developed for each receptor being evaluated (e.g., a scenario-based receptor or simulated individual) in the study area. This type of presentation format shows the aggregate risk for each chemical across multiple pathways, the cumulative risk for each pathway across chemicals, and the overall cumulative cancer risk. In addition, this format allows one to quickly identify both the individual chemicals and pathways that contribute most to the total risk estimate.

	Pathway 1 (Vegetable Ingestion Risk Estimate)[a]	Pathway 2 (Fish Ingestion Risk Estimate)[a]	Pathway 3 (Egg Ingestion Risk Estimate)[a]	Pathway 4 (Beef Ingestion Risk Estimate)[a]	Aggregate Chemical Ingestion Risk Estimate [a]
Exhibit 27. Example Matrix for Estimating Excess Cancer Risks for Multiple Chemical Exposure through Multiple Ingestion Pathways for a Particular Exposure Scenario					
Chemical 1	1×10^{-6}	3×10^{-4}	9×10^{-8}	8×10^{-5}	4×10^{-4}
Chemical 2	4×10^{-7}	4×10^{-6}	4×10^{-8}	4×10^{-7}	5×10^{-6}
Chemical 3	4×10^{-9}	7×10^{-7}	3×10^{-8}	9×10^{-9}	8×10^{-7}
Chemical 4	9×10^{-7}	1×10^{-6}	6×10^{-7}	6×10^{-7}	3×10^{-6}
Cumulative Ingestion Pathway Risk Estimate [a]	3×10^{-6}	3×10^{-4}	7×10^{-7}	8×10^{-5}	4×10^{-4}

[a] Standard rules for rounding apply which will commonly lead to an answer of one significant figure in both risk and hazard estimates. For presentation purposes, hazard quotients (and hazard indices) and cancer risk estimates are usually reported as one significant figure.

4.5.1   Cancer Risk Estimates

Estimated individual cancer risk is expressed as the probability that a person will develop cancer as a result of the estimated exposure over a lifetime. This predicted risk is the **incremental risk** of cancer from the exposure being analyzed (i.e., it does not take into account cancer risk from other factors). Due to default assumptions in their derivation, oral cancer slope factors (CSFs) are generally considered to be "plausible upper-bound" estimates, regardless of whether they are based on statistical upper bounds or best fits. Risks may be estimated for both the central tendency (average exposure) case and for the high end (exposure that is expected to occur in the upper range of the distribution) case, or probabilistic techniques can be used to develop a distribution of estimated risks.

4.5.1.1   Characterizing Individual Pollutant Risk

The first step in characterizing individual pollutant risk for an exposure scenario (e.g., a recreational fisher) is to quantify risk for each ingestion exposure pathway being evaluated. In this step, cancer risks for individual pollutants are estimated by multiplying the estimate of the lifetime average daily dose (LADD) for each ingestion exposure pathway by the appropriate CSF to estimate the potential incremental cancer risk:

$$Risk = LADD \times CSF$$

where:

Risk	=	Individual cancer risk (expressed as an upper-bound risk of contracting cancer over a lifetime) for each pollutant via the ingestion pathway being evaluated (unitless);
LADD	=	Lifetime Average Daily Dietary Intake rate for the pollutant via the ingestion pathway being evaluated (mg/kg-d); and
CSF	=	Oral Cancer Slope Factor for the pollutant via the ingestion pathway being evaluated [(mg/kg-d)$^{-1}$]

Estimates of cancer risk are usually expressed as a probability represented in scientific notation as a negative exponent of 10. For example, an additional upper bound risk of contracting cancer of 1 chance in 10,000 (or one additional person in 10,000) is written as 1×10^{-4}. Because CSFs are typically upper-bound estimates, actual risks may be lower than predicted – note that "the true value of the risk is unknown and may be as low as zero."[25]

In this example approach, risks are evaluated initially for **individuals** within the potentially exposed population. **Population risks** for the exposed population may also be estimated, which may be useful in estimating potential economic costs and benefits from risk reduction. Sensitive subpopulations should also be considered, when possible. Estimates of **incidence** also are possible, although in small populations, even a very high individual risk estimate may not yield an estimated incidence above one case of cancer.

For carcinogens being assessed based on the assumption of nonlinear dose-response, for which an RfD was derived that considers cancer as well as other effects, the hazard quotient approach is appropriate for risk characterization (see below). Where detailed information on carcinogenic mechanisms exists it may also be possible to estimate risk directly from a nonlinear low-dose extrapolation. This approach is supported by EPA's *Guidelines for Carcinogen Risk Assessment*.[11]

4.5.1.2   Characterizing Risk from Exposure to Multiple Pollutants

By each exposure pathway of a scenario, exposure may be to multiple chemicals at the same time rather than a single chemical; however, CSFs are usually available only for individual compounds within a mixture. Consequently, a component-by-component approach is usually employed.[14] The following equation estimates the predicted cumulative incremental individual cancer risk from multiple substances for a single exposure pathway, assuming additive effects from simultaneous exposures to several carcinogens:

$$\text{Risk}_T = \text{Risk}_1 + \text{Risk}_2 + \ldots + \text{Risk}_i$$

where:

Risk_T	=	Cumulative individual ingestion cancer risk (expressed as an upper-bound risk of contracting cancer over a lifetime); and
Risk_i	=	Individual ingestion risk estimate for the ith substance.

In screening-level assessments of carcinogens for which there is an assumption of a linear dose-response, the cancer risks predicted for individual chemicals may be added to estimate cumulative cancer risk for each pathway. This approach is based on an assumption that the risks associated with individual chemicals in the mixture are additive. In more refined assessments, the chemicals being assessed may be evaluated to determine whether effects from multiple chemicals are synergistic (greater than additive) or antagonistic (less than additive), although sufficient data for this evaluation are usually lacking. In those cases where CSFs are available for a chemical mixture of concern, risk characterization can be conducted on the mixture using the same procedures used for a single compound.

For carcinogens being assessed based on the assumption of nonlinear dose-response, for which an RfD considering cancer as well as other effects has been derived, the hazard quotient approach will be appropriate (see Volume 1, Chapter 12).

4.5.1.3   Combining Risk Estimates across Multiple Ingestion Pathways

To evaluate risks associated with the aggregate exposure to a single PB-HAP across multiple pathways of a given scenario, the individual pollutant cancer risk estimates may be summed for each chemical across the multiple ingestion pathways assessed. Additionally, a cumulative multi-pathway risk estimate may be derived by summing cumulative (multiple pollutant) cancer risk estimates across the multiple ingestion pathways.

4.5.1.4   Evaluating Risk Estimates from Inhalation and Ingestion Exposures

Depending on the ingestion scenario, the inhalation pathway may also have been assessed. In such cases, it may be possible to obtain an overall estimate of risk across all pathways by combining the inhalation exposures with the ingestion exposures. It is important to note, however, that the methods and assumptions used to derive the inhalation and ingestion risks may not always yield compatible exposure scenarios. Consequently, it may be sufficient to simply qualitatively consider any potential cumulative risk across routes. This is particularly important when population-level (versus individual) risk estimates are being developed. For example, a scenario-based ingestion exposure assessment will not be easily amenable to producing estimates of numbers of people at different risk levels, while a population-based inhalation assessment may be more appropriate. For example, it would generally not be appropriate to add an inhalation risk that presumes a 70-year exposure duration with an ingestion pathway that presumes a 30-year exposure duration. Any mismatching of exposure durations among pathways in a multipathway assessment should be carefully considered. *For this reason, combining any risk estimates across the ingestion and inhalation pathways is commonly done only by an experienced toxicologist.*

4.5.2   Noncancer Hazard

For noncancer effects (as well as carcinogens being assessed based on the assumption of nonlinear dose-response), ingestion exposure concentrations are compared to RfDs, which are estimates (with uncertainty spanning perhaps an order of magnitude) of a daily oral exposure to the human population (including sensitive subgroups) that is likely to be without an appreciable risk of deleterious noncancer effects during a lifetime.

As with carcinogens, the development of hazard quotients (HQs) for ingestion typically is performed first for individual air toxics, then hazard indices (HIs) may be developed for multiple pollutant exposures, and may be summed across pathways to develop multiple pathway cumulative hazard estimates. An additional step in the multipathway analysis is to evaluate both ingestion and inhalation hazard estimates. These steps are described in separate subsections below.

4.5.2.1   Characterizing Individual Pollutant Hazard

The first step in characterizing individual pollutant hazard for an exposure scenario (e.g., a recreational fisher) is to quantify hazard for each pollutant being evaluated. For ingestion exposures, noncancer hazards are estimated by dividing the estimate of the Average Daily Dose (ADD) by the chronic oral RfD to yield a hazard quotient (HQ) for individual chemicals:

$$HQ = ADD \div RfD$$

where:

HQ	=	The Hazard Quotient for the pollutant via each ingestion pathway being evaluated (unitless);
ADD	=	Estimate of the Average Daily Dietary Intake rate for the pollutant via the ingestion pathway being evaluated (mg/kg-d); and
RfD	=	the corresponding reference dose for the pollutant via the ingestion pathway being evaluated (mg/kg-d)

In screening assessments, which are routinely built around a particular year's estimate of emissions, the chronic exposure estimate may be based on a somewhat shorter than "chronic" exposure concentration estimate (e.g., the average annual concentration estimated by modeling), employing the simplifying assumption of continued similar conditions for a long-term period. A more refined assessment might then include an estimate of exposure derived using information for the full chronic period of exposure time period (e.g., a lifetime or substantial portion of a lifetime, refined emissions estimates over the long-term period).

Based on the definition of the RfD, an HQ less than or equal to one indicates that adverse noncancer effects are not likely to occur. With exposures increasingly greater than the RfD (i.e., HQs increasingly greater than one), the potential for adverse effects increases. However, the HQ should not be interpreted as a probability, because the overall chance of adverse effects does not increase linearly as exposures exceed the RfD.

4.5.2.2   Multiple Pollutant Hazard

Noncancer health effects data are usually available only for individual compounds within a mixture. In these cases, the individual HQs can be summed together to calculate a multi-pollutant hazard index (HI):

$$HI = HQ_1 + HQ_2 + ...+ HQ_i$$

where:

HI	=	Hazard index; and
HQ_i	=	Hazard quotient for the ith air toxic.

For screening-level assessments, a simple HI may first be calculated for all chemicals of concern (Exhibit 28). This approach is based on the assumption that even when individual pollutant levels are lower than the corresponding reference levels, some pollutants may work together such that their potential for harm is additive and the combined exposure to the group of chemicals poses greater likelihood of harm. Some groups of chemicals can also behave antagonistically, such that combined exposure poses less likelihood of harm, or synergistically, such that combined exposure poses harm in greater than additive manner. Where this type of HI exceeds the criterion of interest, a more refined analysis is warranted. However, note that interpretation of differences among HQs across substances may be limited by differences among RfDs in their derivation and the fact that the slope of the dose-response curve above the RfD can vary widely depending on the substance, type of effect, and exposed population.

Exhibit 28. Example Matrix for Characterizing Hazard for Multiple Chemical Exposure through Multiple Ingestion Pathways for a Particular Exposure Scenario					
	Pathway 1 (Vegetable Ingestion HQ)[a]	Pathway 2 (Fish Ingestion HQ)[a]	Pathway 3 (Egg Ingestion HQ)[a]	Pathway 4 (Beef Ingestion HQ)[a]	Aggregate Chemical Ingestion HQ [a]
Chemical 1	2×10^{-1}	2×10^{-1}	4×10^{-2}	2×10^{-1}	7×10^{-1}
Chemical 2	3×10^{-1}	7×10^{-1}	3×10^{-2}	2×10^{-1}	1
Chemical 3	1×10^{-1}	4×10^{-1}	2×10^{-1}	4×10^{-1}	1
Chemical 4	9×10^{-2}	1×10^{-2}	1×10^{-1}	2×10^{-2}	3×10^{-1}
Cumulative Ingestion Pathway HI [a]	7×10^{-1}	1	4×10^{-1}	9×10^{-1}	3

[a] Standard rules for rounding apply which will commonly lead to an answer of one significant figure in both risk and hazard estimates. For presentation purposes, hazard quotients (and hazard indices) and cancer risk estimates are usually reported as one significant figure.

The assumption of dose additivity is most appropriate to compounds that induce the same effect by similar modes of action. Thus, EPA guidance for chemical mixtures[14] suggests subgrouping pollutant-specific HQs by toxicological similarity of the pollutants for subsequent calculations; that is, to calculate a target-organ-specific-hazard index (TOSHI) for each subgrouping of pollutants. This calculation allows for a more appropriate estimate of overall hazard.

The HI approach encompassing all chemicals in a mixture may be appropriate for a screening-level study. However, it is important to note that applying the HI equation to compounds that may produce different effects, or that act by different mechanisms, could overestimate the potential for effects. Consequently, in a refined assessment, it is more appropriate to calculate a separate HI for each noncancer endpoint of concern when mechanisms of action are known to be similar.

4.5.2.3   Evaluating Hazard Estimates From Inhalation and Ingestion Exposures

As with carcinogenic assessments, in some cases it may be possible to combine estimates of inhalation hazard with ingestion hazard estimates to provide an estimate of total hazard across all exposure pathways for a receptor. As noted earlier, this is a complex analysis that generally is conducted by an experienced toxicologist.

4.5.3   Consideration of Long-Range Transport and Background

Although the example approach presented here focuses on populations located close to a facility/source, the contribution of a facility/source to deposition of PB-HAPs in more distant environmentally sensitive areas (e.g., the Great Lakes, Chesapeake Bay, the Florida Everglades) through long-range transport may also be a potential concern. Even small contributions from a single facility may become significant over time. In addition, while local residents may not eat fish from bodies of water around a facility/source, fish caught from the large, productive lakes and bays in the U.S. are known to be eaten by significant numbers of people. Risk assessors are encouraged to discuss the contribution by an evaluated facility/source to the potential risks associated with this exposure route.

Similarly, the study area around a specific facility/source may be subject to deposition of PB-HAPs resulting from long-range transport from more distant sources. Risk assessors are encouraged to discuss the contribution of these background levels of deposition as appropriate.

Note, however, that the local impacts of PB-HAP emissions may be of greatest potential concern for many facilities/sources. Therefore, whether or not to include a detailed assessment of long-range transport and/or background generally is an analysis-specific decision (often determined by legal and regulatory requirements).

4.5.4   Assessment and Presentation of Uncertainty

In the final part of this example risk characterization, estimates of cancer risk and noncancer hazard are presented in the context of uncertainties and limitations in the data and methodology. Exposure estimates and assumptions, toxicity estimates and assumptions, and the assessment of uncertainty and variability commonly are discussed.

The risk estimates used in multipathway toxics risk assessments are unlikely to be fully probabilistic estimates of risk because of the numerous pathways, potential receptors, and potential parameter values involved in the assessment. Rather, such risk estimates are likely to be conditional estimates that incorporate a considerable number of assumptions about exposure and toxicity. Multipathway air toxics risk assessments are subject to additional sources of uncertainty and variability as compared to inhalation risk assessments. The multimedia modeling effort is both more complex and less certain due to many factors. For example: (1) there are many more chemical-dependent and chemical-independent variables involved as input values to the models; (2) the models involve analysis of the transfer of air toxics from the air to other media (e.g., soil, sediment, water), the subsequent movement of the air toxics between these media (e.g., soil runoff to surface water), and uptake and metabolism by biota; and (3) many variables affect the ingestion of food, water, and other media by humans and wildlife, and the exposure and risk estimates may differ considerably as a consequence of the assumptions used to derive intake estimates. Sampling of biota and abiotic media also may be more complex and uncertain. Additional uncertainties are incorporated in the risk assessment when exposure estimates to multiple substances across multiple pathways are summed.

As a result of the increased complexity and uncertainty of the analytical approach, multipathway risk assessments commonly include semi-quantitative sensitivity analyses and may include quantitative uncertainty analysis. These approaches to uncertainty analysis are discussed in greater detail in Section III-5.4 and in Volume 1 (Chapters 3 and 13) of this reference library.

4.6   Tier 1 Multipathway Human Health Analysis

(reserved)

4.7   Tier 2 Multipathway Human Health Analysis

4.7.1   Introduction

This section describes an example approach for performing a Tier 2 multipathway risk assessment. Exhibit 29 provides an overview of this example approach. In this example approach, ISCST3 is used to provide deposition rates to soils and surface waters. The MPE methodology is used to estimate human exposure at the most impacted location via several ingestion exposure pathways. Because EPA has not yet identified a Tier 1 approach, this example Tier 2 analysis incorporates simplified assumptions that allow a conservative risk/hazard estimate to be calculated with a relatively modest analytical effort. If the facility/source passes this screening analysis, the risk manager can be reasonably confident that the likelihood for significant risk/hazard is low. As a consequence, monitoring is not explicitly included in this example Tier 2 analysis (although available monitoring data can be used as inputs to the exposure analysis).

4.7.2   Fate and Transport Modeling

This example Tier 2 approach uses the ISCST3 model for estimating air concentrations for both chronic and acute exposures (an alternative would be to use AERMOD). Other models may be relevant for a specific facility/source. For example, if the facility/source is located in complex terrain and/or near a large body of water, other models may be more applicable (e.g., Cal-Puff for complex terrain; non-Gaussian models for sources close to large bodies of water). Similarly, if the exposure point of concern is very near source (< 100-m) or very distant (> 50-km), other models would need to be applied (e.g., CALPUFF for > 50-km). However, where the assessment could support a regulatory decision, the use of an alternative model commonly is agreed to in advance with the regulatory agency decision-maker. Alternative models that conform to EPA's air quality modeling guidance[17] are more likely to be acceptable to those decision-makers.

4.7.2.1   Model Inputs

The modeling for the multipathway analysis commonly is executed in about the same way as would be done for a Tier 2 inhalation analysis. For example, sources are characterized the same way regarding stack/vent height and diameter, release temperature and velocity, flow rate, etc. The ISCST3 model requires the speciation profile of the emissions for ISCST3 in order to calculate deposition rates properly. The risk assessor would have addressed deposition and these properties in the Tier 2 inhalation analysis if he/she executed the ISCST3 inhalation runs "with depletion" (i.e., telling the model to subtract out the mass of chemical deposited). This is a concern for emissions of chemicals that are predominantly particulates instead of vapors and/or are predominantly bound to particles (e.g., metals). However, *the inhalation runs may not have been done with depletion*, because it takes longer to do this. If the inhalation runs were done without depletion, they should be done again with depletion for the multipathway analysis.

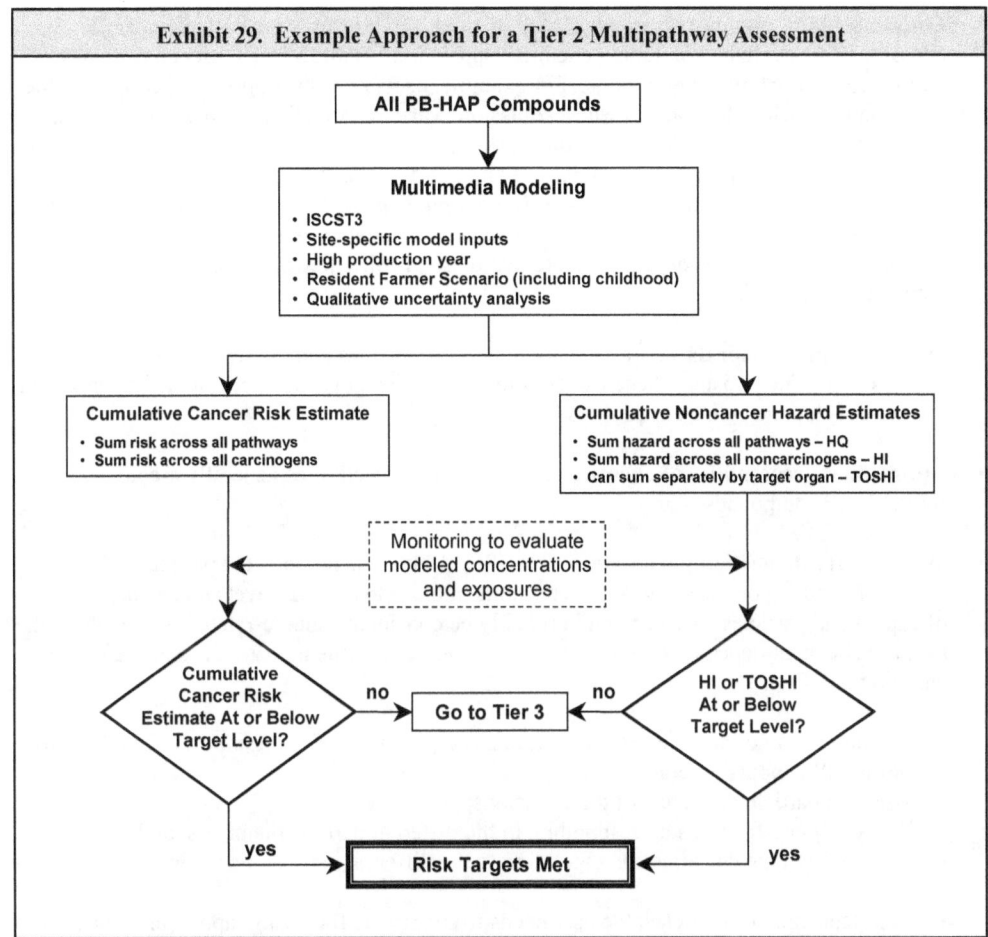

Exhibit 29. Example Approach for a Tier 2 Multipathway Assessment

All PB-HAP Compounds

Multimedia Modeling
- ISCST3
- Site-specific model inputs
- High production year
- Resident Farmer Scenario (including childhood)
- Qualitative uncertainty analysis

Cumulative Cancer Risk Estimate
- Sum risk across all pathways
- Sum risk across all carcinogens

Cumulative Noncancer Hazard Estimates
- Sum hazard across all pathways – HQ
- Sum hazard across all noncarcinogens – HI
- Can sum separately by target organ – TOSHI

Monitoring to evaluate modeled concentrations and exposures

Cumulative Cancer Risk Estimate At or Below Target Level?

Go to Tier 3

HI or TOSHI At or Below Target Level?

no → ← no

yes → **Risk Targets Met** ← yes

Inputs for ISCST3

Input data for ISCST3 fall into seven general categories: (1) source location; (2) emissions data; (3) stack/vent parameter data; (4) pollutant-specific data (reactivity); (5) wet and dry deposition parameters; (6) meteorological data; and (7) population data. The ISC user's guide[9] provides more detailed information on the deposition algorithms and required input data. There also is guidance for application of ISC for multipathway assessment in the latest MPE documentation.[19] *This example approach assumes that all input data are at a level of detail equivalent to the Tier 2 inhalation analysis.*

- **Source location**. ISCST3 requires coordinate data (latitude and longitude) for each emission source in the analysis. Multiple sources can be modeled in the same ISCST3 run.

- **Emissions data**. In this example approach, the risk assessor uses the best site-specific emissions data available, including chemical speciation. With ISCST3, users have many options for characterizing emissions. For example, users have the option to specify variable emission rate factors for sources whose emissions vary as a function of time (e.g., month, season, hour-of-day). In addition, settling velocity categories, mass fractions, and reflection coefficients may be specified for sources of large particulates that experience settling and removal during dispersion. *Therefore, this example approach assumes that the emissions profile(s) used for modeling reflect the expected pattern(s) of emissions over a reasonable period of time (e.g., several years)*. Note that these profiles may differ for different sources within a single facility.

- **Stack/vent parameter data**. Facility/source-specific values commonly are used for all stack/vent parameter data. These values commonly reflect the expected patterns of emissions used to develop the emissions profiles for modeling.

- **Atmospheric reactivities of chemicals**. Atmospheric reactivities generally are not needed for PB-HAP compounds.

- **Wet and dry deposition parameters**. The ISCST3 models requires the particulate/ particle-bound/vapor fractions of the emissions in order to calculate wet and dry deposition of vapors and particles. These would probably be considered source-related, since although they are chemical-dependent, they also vary by source (i.e., the industrial process affects the emissions profile).

  For dry deposition of particles, the user needs to supply the following inputs (in addition to the normal ISC inputs), including the:
  – Array of particle diameters of the emissions;
  – Array of mass fractions corresponding to the different particle diameters; and
  – Array of particle densities corresponding to the different particle diameters.

  For wet deposition of particles, the user needs to supply the following inputs (in addition to the normal ISC inputs), including the:
  – Particle scavenging coefficients for liquid precipitation corresponding to the different particle diameters; and
  – Particle scavenging coefficients for frozen precipitation corresponding to the different particle diameters.

  For wet deposition of gases, the user needs to supply the following inputs (in addition to the normal ISC inputs), including the:
  – Gaseous scavenging coefficient for liquid precipitation; and
  – Gaseous scavenging coefficient for frozen precipitation.

- **Meteorology**. Tier 2 multipathway analyses commonly use the most recent consecutive five years of facility/source-specific data from the nearest representative meteorological station. "Representative" generally means that the sources being modeled and the weather station are located in the same general environment with respect to *significant terrain* (e.g., valley vs. plateau), *significant geographic features* (e.g., proximity to a large body of water), and *prevailing winds* (i.e., similar wind rose/direction of dominant wind). Documentation demonstrating the representativeness of the meteorological data is encouraged. Sources of meteorological data are provided in Volume 1, Appendix G.

- **Population**. Population is not defined for the ISCST3 model runs (population is defined using MPE (see Section 4.7.3 below).

Inputs for the MPE Methodology

The MPE methodology requires air concentrations and deposition rates for the appropriate locations as inputs. These are provided by ISCST3. A variety of additional chemical- and/or facility/source-specific parameters are needed for MPE to calculate concentrations in various media (e.g., soil, surface water) and biota (e.g., plants, fish). These are identified and described in detail in the MPE documentation.[19]

4.7.2.2   Model Runs

Using one complete year of meteorological data, ISCST3 calculates annual deposition rates (or fluxes) for each receptor location specified in a run. Results for each of the five years of meteorological data are obtained using separate ISCST3 model runs for each year. ISCST3 can model all sources at the facility simultaneously; however, only one chemical can be modeled at a time. Therefore, for each PB-HAP compound, all sources of that compound could be modeled in the same ISCST3 model run.

Defining receptor locations

For this example Tier 2 analysis, the subsistence farmer is assumed to occur at the modeling location ("node") with the maximum deposition rate. Several iterations of the ISCST3 model may be required to locate this point. The risk assessor is encouraged to clearly document the process used to identify this point.

- If multiple PB-HAPs are present, the point where maximum deposition occurs for each individual PB-HAP may not be in the same location. This example approach assumes that these are all co-located (see Section 4.7.4 below).

- For this example approach, the point where maximum deposition occurs is used for all soil pathways and for the drinking water pathway.

- For surface water pathways, this example Tier 2 analysis is based on the surface water body within the assessment area where maximum concentrations occur. This will depend on the location of the surface water bodies with respect to sources, characteristics of the surface water body (e.g., size, depth, flow rate), and characteristics of the watershed (e.g., size, runoff, erosion potential). Several water bodies/watersheds within the assessment area may need to be characterized in order to identify this water body. The risk assessor is encouraged to clearly document the process used to identify this water body.

- For this example approach, the subsistence farmer is assumed to fish in the water body where maximum concentrations occur.

Defining the duration of emissions

Because the concentrations of PB-HAP compounds often will slowly accumulate in soil, sediment, and biota over time, the total deposition over time (and the resulting media concentrations) will depend on the specific **duration of emissions** selected for the analysis.

For a Tier 2 analysis, the duration of emissions used in the analysis commonly reflects the expected duration of emissions from the facility/source being evaluated. A common duration is 30 or 40 years (e.g., the expected lifespan of many facilities or processes). However, facilities may continue to operate beyond their expected lifespan. Therefore, *the risk assessor is encouraged to document the rationale for selecting a duration of emissions*. In the absence of clear rationale for another value, a conservative value (e.g., 100 years) can be selected.

Defining deposition/flux rates for estimating media concentrations

This example approach uses the average deposition/flux rates from the five years of ISCST3 modeling and assumes these rates are constant over the duration of emissions.

Model outputs to use

The modeled deposition rates and duration of deposition are used as inputs to the MPE methodology to estimate PB-HAP concentrations in soil, water, and biota. This example approach uses the average annual deposition rates as inputs to the MPE methodology.

4.7.3   Exposure Assessment

For this example Tier 2 assessment, the exposure assessment includes both central tendency and high end exposure estimates.

4.7.3.1   Characterization of the Study Population

For this example Tier 2 assessment, the study population is limited to a hypothetical subsistence farmer assumed to reside at the point of maximum deposition and to fish in the water body with the highest modeled concentrations. For assessments for which information of local farming practices is available, the location of actual farms can be used as the basis for a more realistic exposure scenario.

4.7.3.2   Defining the Point of Maximum Exposure

This example Tier 2 assessment defines the point of maximum exposure as follows:

- For a single PB-HAP compound, the point of maximum exposure is defined as the modeled location (node) at which the highest deposition rate occurs (soils) and the surface water body with maximum modeled concentrations. Exposure estimates from these two locations are summed to calculate total maximum exposure.

- For multiple PB-HAP compounds, the point of maximum exposure is defined by assuming that the point of maximum exposure occurs at the same location for each PB-HAP compound (i.e., exposure for each compound is added to calculate total exposure).

4.7.3.3 Defining the Exposure Scenario

This example Tier 2 assessment is based on a single exposure scenario designed to provide a conservative estimate of risk/hazard via all ingestion pathways. This scenario is termed the **subsistence farmer**. This scenario reflects an individual living on a farm and consuming meat, dairy products, and vegetables that the farm produces. All drinking water is obtained from a surface water body on the farm (or if no surface water is present, from collected rainfall). The animals raised on the farm subsist primarily on forage that is grown on the farm. This scenario also assumes that the farm family fishes in the surface water body with the highest concentrations within the study area. They fish at a recreational level and eat the fish they catch. The maximum exposed individual is assumed to be exposed to PB-HAP compounds through the following exposure pathways:

- Direct inhalation of vapors and particles (assessed during the inhalation analysis);
- Incidental ingestion of soil and house dust;
- Ingestion of drinking water from surface water sources;
- Ingestion of homegrown produce;
- Ingestion of home-produced meat, milk, and eggs; and
- Ingestion of fish; and
- Ingestion of breast milk (evaluated separately for an infant [for PCBs, dioxins, and furans]).

Within this general scenario, the specific exposure routes evaluated for a given facility/source may depend on the particular PB-HAP compounds in the emissions being assessed (see Exhibit 26).

4.7.3.4 Calculation of Exposure Concentration

This example Tier 2 approach calculates exposure concentration as follows:

- Exposure concentrations in soil, drinking water, and food items are calculated using the equations provided in the MPE methodology[19] or the HHRAP.[20]

- The maximum concentration (via each exposure pathway) reached during the modeling period is used as the exposure concentration. For a constant emissions scenario, this usually will be the final year of the modeling simulation (e.g., the last year the facility/source is expected to operate).

- Note that existing monitoring data may be used to evaluate or refine the exposure estimates based on multimedia modeling.

4.7.3.5   Determining Exposure

This example Tier 2 approach assumes that the subsistence farmer is exposed to this concentration during his/her entire lifetime, from birth to the age of 70 years. This entails the use of variable exposure factors (e.g., body weight, consumption rate) as the subsistence farmer ages from birth to age 70. Additionally, for the derivation of chronic hazard estimates, a child (e.g., 0-7 years old) may be included to address the potential for substantially higher pollutant intake (in mg/kg-day) during those ages.

4.7.3.6   Determining Intake

For this example approach, the risk assessor calculates a central tendency and high end intake. EPA's *Exposure Factors Handbook*[26] provides default central tendency and high-end values for intake and other exposure factors (e.g. body weight). Equations used to calculate intake are found in the MPE methodology.[19] EPA's *Guidance on Selecting the Appropriate Age Groups for Assessing Childhood Exposures to Environmental Contaminants*[27] and *Child-Specific Exposure Factors Handbook*[28] provide guidance on selecting the appropriate age groups to include for children and the exposure factors to use that are specific to each age group.

4.7.4   Risk Characterization

Risk characterization for this example Tier 2 approach is limited to estimating ingestion risk for the subsistence farmer (calculated separately for cancer risk and chronic noncancer hazard). The estimates of intake rates described in Section 4.7.3.6 are used to calculate risk and hazard according to the basic equations presented earlier in Section 4.5. Background concentrations are not explicitly considered.

As noted earlier, the points of maximum deposition rates for multiple PB-HAPs are assumed to be co-located, and the subsistence farm family is assumed to eat fish from the surface water body with the highest modeled concentrations.

4.7.4.1   Reporting Results

A relatively simple summary can be used to report results, as long as it is consistent with the need to make the results both transparent and reproducible. Examples of reports prepared for EPA's purposes can be found in EPA's residual risk "test memos;" these may or may not be appropriate examples for the specific purposes of other risk assessments. The summary generally will include the following information:

- Documentation of input parameters, output values, and risk characterization, with special emphasis on comparing estimated risk/hazard to risk targets;

- A simple presentation describing the assessment's purpose (e.g., to determine whether risk is below levels of concern) and the outcome relative to that purpose (e.g., low risk is not demonstrated); and

- Documentation of anything that is discretionary (i.e., anything that is facility-specific), such as emissions characteristics or choice of a meteorological station other than the nearest.

4.7.4.2   Assessment and Presentation of Uncertainty

Risk managers need to understand the strengths and the limitations of the Tier 2 multipathway risk assessment. A critical part of the risk characterization process, therefore, is an evaluation of the assumptions, limitations, and uncertainties inherent in the Tier 2 assessment in order to place the risk estimates in proper perspective.[3] Tier 2 assessments commonly include a quantitative or qualitative description of the uncertainty for each parameter and indicating the possible influence of these uncertainties on the final risk estimates given knowledge of the models used. Tier 2 assessments also may include a semi-quantitative sensitivity analyses. These approaches are described in Section III-5.4. Sensitivity analyses are discussed in more detail in Volume 1 (Chapters 3 and 13) of this reference library.

4.7.5   Potential Refinements of a Tier 2 Approach

If the results of the Tier 2 assessment indicate that risk targets are not met, the risk assessor has the option of conducting a much more facility/source-specific, refined Tier 2 assessment using the ISCST3 model and MPE methodology. Refinements might include characterizing the modeling region in much greater detail and using exposure scenarios that are more realistic for the facility/source setting (the exposure assessment for each scenario could still be based on the basic equations found in the MPE methodology). Potential refinements might include:

- Defining and assessing a "resident" scenario based on modeled deposition rates at locations where people actually live (e.g., census tract internal points);

- For the "subsistence farmer," (a) use an actual location of a farm within the study area (rather than assuming it is located at the point of maximum deposition), (b) more realistically incorporate only a subset of the potential pathways, and/or (c) assume that the farmer obtains some of his/her food from other sources.

- When summing exposures over multiple pathways, use high-end concentrations for some pathways and central tendency concentrations for others.

- Incorporating a limited monitoring program to evaluate key exposure estimates (e.g., concentrations of PB-HAPs in various food items).

A more complete listing of potential exposure scenarios is presented in Volume 1 (Part III) of this reference library. EPA's *Guidelines for Exposure Assessment*[29] discusses how to develop these types of exposure scenarios.

4.8   Tier 3 Multipathway Human Health Analysis

Note that the discussion in this section is based on information about the design of TRIM.Expo$_{Ingestion}$ and might need to be modified in future versions of this document.

4.8.1   Introduction

This section describes an example approach for performing a Tier 3 multipathway risk assessment. Exhibit 30 provides an overview of this example approach. This example Tier 3 assessment is significantly different than the Tier 2 example approach, in that it involves the use of TRIM.FaTE for multimedia modeling, specific consideration of population locations, and the use of an exposure model (TRIM.Expo$_{Ingestion}$).

This example Tier 3 analysis allows considerable flexibility in analytical approach and detail. For example:

- TRIM.FaTE is used for multimedia modeling. This component of the TRIM modeling system accounts for the movement of a chemical through a comprehensive system of discrete compartments (e.g., media and biota) that represent possible locations of the chemical in the physical and biological environments of the modeled ecosystem and provides an inventory, over time, of a chemical throughout the entire system. TRIM.FaTE generates media concentrations relevant to human ingestion exposures that can be used as input to TRIM.Expo$_{Ingestion}$. TRIM.FaTE allows considerable spatial refinement in selecting exposure locations, and provides concentrations for user-specified (actual) locations.

- TRIM.Expo$_{Ingestion}$ is used to estimate the ingestion intake rates used for the risk characterization. Estimated media and fish concentrations needed by TRIM.Expo$_{Ingestion}$ are provided by TRIM.FaTE. A farm food chain model (employing MPE equations[19]) will also be available to provide livestock and produce contaminant estimates from air and soil concentrations and air deposition estimates provided by TRIM.FaTE or from an external file.

- A risk model (TRIM.Risk$_{HH}$) is used to calculate cumulative excess cancer risk and noncancer hazard associated with the modeled dietary intake estimates. TRIM.Risk$_{HH}$ calculates human health risk metrics, documents model inputs and assumptions, and displays results.

- Monitoring data may be incorporated into the exposure assessment as inputs to TRIM.Risk$_{HH}$.

This example Tier 3 analysis is highly facility/source-specific and requires careful planning. Where the assessment could support a regulatory decision, advance discussions with the regulatory agency risk manager are strongly encouraged.

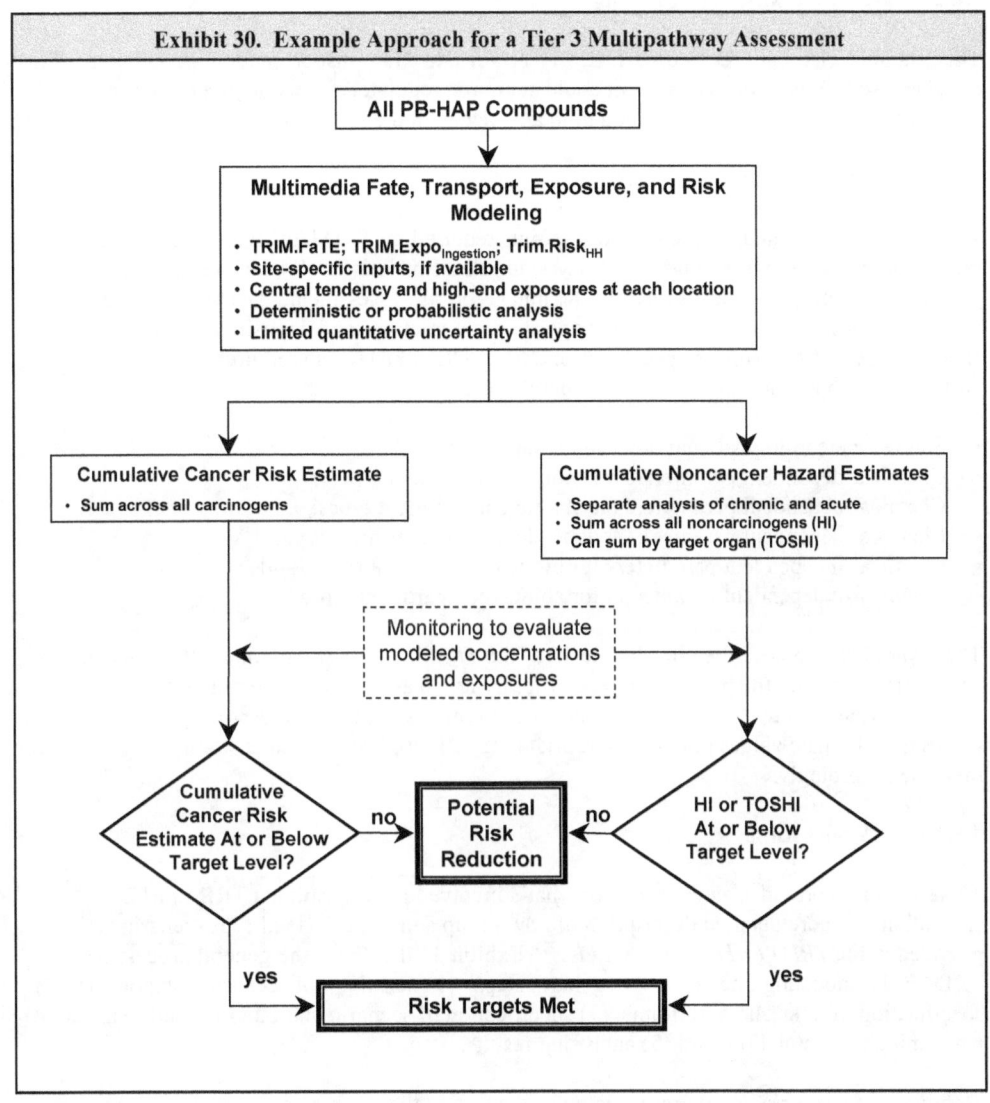

Exhibit 30. Example Approach for a Tier 3 Multipathway Assessment

All PB-HAP Compounds

Multimedia Fate, Transport, Exposure, and Risk Modeling
- TRIM.FaTE; TRIM.Expo$_{Ingestion}$; Trim.Risk$_{HH}$
- Site-specific inputs, if available
- Central tendency and high-end exposures at each location
- Deterministic or probabilistic analysis
- Limited quantitative uncertainty analysis

Cumulative Cancer Risk Estimate
- Sum across all carcinogens

Cumulative Noncancer Hazard Estimates
- Separate analysis of chronic and acute
- Sum across all noncarcinogens (HI)
- Can sum by target organ (TOSHI)

Monitoring to evaluate modeled concentrations and exposures

Cumulative Cancer Risk Estimate At or Below Target Level?

HI or TOSHI At or Below Target Level?

no   **Potential Risk Reduction**   no

yes   **Risk Targets Met**   yes

4.8.2 Fate and Transport Modeling

This example Tier 3 approach uses TRIM.FaTE for fate and transport modeling. Other models could be used. Where the assessment could support a regulatory decision, the use of an alternative model commonly is agreed to in advance with the regulatory agency decision-maker.

4.8.2.1 Model Inputs

The only facility-related/source term data points required by TRIM.FaTE are chemical emission rate, location (i.e., latitude/longitude, UTM), and emission height, all of which should be available from modeling performed for the inhalation risk assessment. TRIM.FaTE does all the calculations internally for determining vapor/particle fractions and deposition rates based on chemical-specific (vs. source-specific) properties. TRIM.FaTE also requires a number of input variables which fall into the following general categories:

- Source, meteorological, and other input parameters;
- Chemical-dependent parameters for biotic compartment types;
- Chemical-dependent parameters for abiotic compartment types;
- Chemical-dependent parameters independent of compartment type;
- Chemical-independent parameters for biotic compartment types; and
- Chemical-independent parameters for abiotic compartment types.

These variables are described in Module 16 of the *TRIM.FaTE User's Guide*.[21] Note that the values provided with the public Reference Library are to assist users in learning how to use TRIM.FaTE and set up TRIM.FaTE scenarios. It remains each user's responsibility to confirm or identify alternate values that are appropriate for their application and customize the library for their use accordingly.

4.8.2.2 Model Runs

This subsection provides an overview of what's involved in performing a TRIM.FaTE simulation. A more detailed description of how set up and run a TRIM.FaTE scenario is provided in the *TRIM.FaTE User's Guide*.[21] Exhibit 31 illustrates the general process for TRIM.FaTE modeling, including five general steps: (1) definition of scenario components; (2) specification of links and algorithms; (3) specification of scenario and other property values; (4) performing the simulation; and (5) analyzing results.

Definition of scenario components includes identifying the chemicals and sources to be included (i.e., the facility/sources and PB-HAPs), determining the modeling region, and specifying parcels, volume elements, and compartment types. Information to assist TRIM.FaTE users in all aspects of designing the scenario is provided in the *TRIM.FaTE User's Guide*.[21] Additionally, the user will need to consider facility/source-specific information sources for each application (e.g., National Land Cover Data at http://www.epa.gov/mrlc/nlcd.html - for parcel set up).

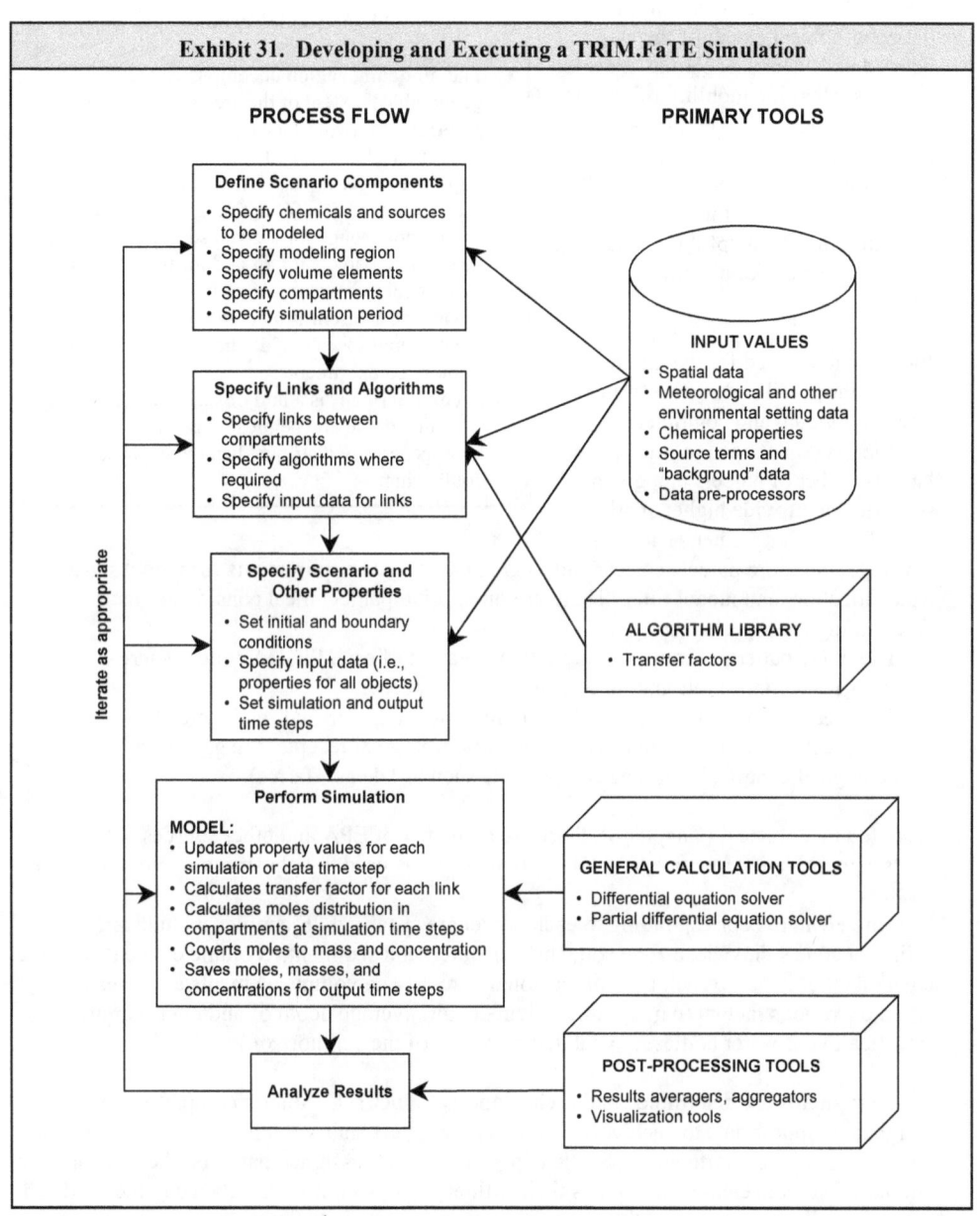

Exhibit 31. Developing and Executing a TRIM.FaTE Simulation

PROCESS FLOW

PRIMARY TOOLS

Define Scenario Components
- Specify chemicals and sources to be modeled
- Specify modeling region
- Specify volume elements
- Specify compartments
- Specify simulation period

Specify Links and Algorithms
- Specify links between compartments
- Specify algorithms where required
- Specify input data for links

Specify Scenario and Other Properties
- Set initial and boundary conditions
- Specify input data (i.e., properties for all objects)
- Set simulation and output time steps

Perform Simulation
MODEL:
- Updates property values for each simulation or data time step
- Calculates transfer factor for each link
- Calculates moles distribution in compartments at simulation time steps
- Coverts moles to mass and concentration
- Saves moles, masses, and concentrations for output time steps

Analyze Results

INPUT VALUES
- Spatial data
- Meteorological and other environmental setting data
- Chemical properties
- Source terms and "background" data
- Data pre-processors

ALGORITHM LIBRARY
- Transfer factors

GENERAL CALCULATION TOOLS
- Differential equation solver
- Partial differential equation solver

POST-PROCESSING TOOLS
- Results averagers, aggregators
- Visualization tools

Iterate as appropriate

- The **modeling region** encompasses the geographical extent of the area to be modeled. The user should consider factors such as the mobility of the PB-HAPs, location of the facility/source, location of sensitive populations, and background concentrations of the PB-HAPs. The results of inhalation modeling may be helpful in evaluating the predicted spatial pattern of deposition.

- **Parcels** are defined for the air and land surface (soil and surface water). Their individual and combined boundaries do not need to line up. A larger number of parcels in a given scenario can provide higher spatial resolution and/or greater aerial coverage, but more parcels correspond to greater resource requirements for model set-up, data collection, and model runs. There are three principal technical considerations for determining the parcels:
  - The likely pattern of transport and transformation of each PB-HAP (i.e., where significant concentration gradients are likely to occur);
  - The locations of natural and land use boundaries (e.g., airsheds and watersheds); and
  - The locations of important environmental or biological receptors (e.g., human or ecological cohorts, landscape components such as lakes or farms).

As noted in Volume 1 (Chapter 6), there are a number of EPA and other sources of landscape, elevation, climate, and GIS data that can be used to help structure modeling runs.

- **Volume elements** corresponding to each parcel are specified. Examples include air, surface soil, root zone soil, vadose zone soil, surface water, and sediment. Volume elements add the component of depth to each two-dimensional parcel. The volume elements are determined based on various factors (e.g., mixing heights in air, average depth of and approximate stratification of water bodies, typical demarcations of the soil horizon).

- **Compartments** are determined and their property values set. Abiotic compartments are assigned as appropriate to each volume element (e.g., each air volume element is assigned one or more air compartments). Biotic compartments are assigned based on their occurrence within the volume element as well as their influence on overall mass balance in the modeled scenario and concentration in the compartments of interest in the assessment. The user has flexibility in assigning biotic compartments (e.g., a single species can represent an entire trophic group, or the user can specify a particular species of concern such as an endangered species).

> **TRIM.FaTE Terminology**
>
> - The **modeling region** encompasses the geographical extent of the area to be modeled.
> - A **parcel** is a two-dimensional, horizontal geographical area used to subdivide the modeling region. Parcels can be virtually any size and shape, are the basis for defining volume elements, and do not change for a given scenario. There can be separate parcels for air and for the land surface (soil or surface water).
> - A **volume element** is a bounded, three-dimensional space that defines the location of one or more compartments.
> - A **compartment** is a unit of space within which it is assumed that all chemical mass is homogeneously distributed and is in phase equilibrium.

- The **simulation period** is determined. As noted with the Tier 2 analysis, this commonly reflects the expected duration of emissions from the facility/source being evaluated. Note, however, that the rate at which a modeled chemical accumulates in all compartments slows with time, and for chemicals that degrade there may likely be a leveling off of concentration by 30-40 years. Therefore, the duration selected also should reflect the time necessary for the rates of accumulation in the media/biota of interest to slow to a negligible rate of increase or cease. If information on accumulation rate is not available, the steady-state solution may be an acceptable output. Some additional thought is required for this option during set-up, but run time and the time needed to process results is reduced.

Specification of links and algorithms is the step in which links between compartments and sinks are specified for a given scenario. Generally, adjacent abiotic compartments are linked to each other, and compartments at the edge of the modeling region are linked to advection sinks. Biotic compartments generally need to be linked to the appropriate abiotic and biotic compartments. For example, each biotic compartment should be linked to all compartments that comprise its diet or in any way provide it a source of chemical mass and all to which it passes mass. The system of links is very important, because the links provide for the assignment of the algorithms describing the processes that drive chemical transfer and transformation (e.g., with the current set of library algorithms, TRIM.FaTE cannot simulate bioconcentration in benthic invertebrates if that biotic compartment is not linked to the abiotic sediment compartment).

Specifying the scenario and other properties involves providing the remaining relevant input values needed for the simulation. This involves specifying the chemical properties of each PB-HAP being modeled, any initial distribution of chemical mass in the compartments, the data for each modeled source, environmental data needed by the selected algorithms, and the simulation and output time steps. A list of all the parameters associated with the current Reference Library algorithms, which, to date, have been set to constant or time-varying numeric values (e.g., as compared to equations or functions) is presented in Appendix D of Volume II of the *TRIM.FaTE Technical Support Document* (http://www.epa.gov/ttn/fera/trim_fate.html).

- **Chemical-specific properties** include molecular weight, melting point, half-life or degradation rates, chemical transformation rates, and elimination rates.

- **Initial and boundary conditions** (e.g., initial concentrations in specific media) can be specified.

- **Source data** include location, height, and emission rate.

- **Environmental setting data** include meteorological data (at any or various time intervals), data needed to define the characteristics of the biotic and abiotic compartments being evaluated (e.g., soil density, current flow velocity, biomass per area, food ingestion rate), and data needed to define properties of links between compartments (e.g., fraction of total surface runoff from a soil compartment to an adjacent water compartment). Collecting these data may be difficult and time-consuming.

- The **simulation time step** specifies a minimum frequency at which the model will calculate transfer factors and chemical mass exchange between (and transformed within) compartments. The **output time step** determines the points in time at which the amount of chemical in each compartment will be reported as output as moles, mass, or concentration.

Performing the simulation is the actual running of the model.

Analyzing results is facilitated with several TRIM.FaTE tools:

- The **Averager** can generate averages of TRIM.FaTE outputs in any multiple of the output time step as well as in monthly and annual increments. It can also limit the compartments included in the averaged file.

- The **Graphical Results Viewer** presents model results (moles, mass, or concentration) on a map of the parcels by using different colors to represent incremental gradients in the results for a specific chemical.

- The **Aggregator** can produce tables in HTML, text, or comma-delimited formats that combine columns of output data in different ways for producing combined or comparative statistics. Functions available for combining results include sum, average, difference, ratio, and percent difference.

4.8.3   Exposure Assessment

This example Tier 3 exposure assessment requires careful planning. Where the assessment could support a regulatory decision, advance discussions with the regulatory agency risk manager are encouraged. TRIM.Expo$_{Ingestion}$ provides a more flexible way to assess exposure scenarios for multipathway assessments.

> If TRIM.Expo$_{Ingestion}$ is not available, risk assessors can use TRIM.FaTE to calculate PB-HAP concentrations in media and biota and then use the basic intake equations in the MPE methodology to estimate exposure for each specific scenario evaluated.

The initial version of TRIM.Expo$_{Ingestion}$ incorporates a scenario-based approach to exposure assessment. It calculates dietary intake using the output from TRIM.FaTE modeling (e.g., concentrations) to calculate ingestion intake using the generic intake equation presented in Section 4.3 above. TRIM.Expo$_{Ingestion}$ currently supports 13 exposure scenarios (Exhibit 32).

Within each scenario, TRIM.Expo$_{Ingestion}$ provides the flexibility to incorporate a variety of assumptions regarding exposure duration (ED), exposure frequency (EF), and varying age groups within the ED period (e.g., an individual exposed as both a child and an adult).

Exhibit 32. Scenarios[a] Currently Supported by TRIM.Expo$_{Ingestion}$

- **Resident**: Ingestion of soil and water
- **Resident Gardener**: Ingestion of soil, water, home-grown fruits, vegetables and grains
- **Resident Fisher**: Ingestion of soil, water, local fish
- **Resident Gardener Fisher**: Ingestion of soil, water, home-grown fruits, vegetables and grains, and local fish
- **Beef Farmer Fisher**: Ingestion of soil, water, locally-produced fruits, vegetables and grains, locally-produced beef, and local fish
- **Dairy Farmer Fisher**: Ingestion of soil, water, locally-produced fruits, vegetables and grains, locally-produced dairy, and local fish
- **Beef Farmer**: Ingestion of soil, water, locally-produced fruits, vegetables and grains, locally-produced beef
- **Dairy Farmer**: Ingestion of soil, water, locally-produced fruits, vegetables and grains, locally-produced dairy
- **Poultry Farmer**: Ingestion of soil, water, locally-produced fruits, vegetables and grains, locally-produced poultry
- **Egg Farmer**: Ingestion of soil, water, locally-produced fruits, vegetables and grains, locally-produced eggs
- **Pork Farmer**: Ingestion of soil, water, locally-produced fruits, vegetables and grains, locally-produced pork
- **Subsistence Beef Farmer**: Same as beef farmer but fractions contaminated for fruits, vegetables, grains, and beef is set to 1 rather than applying the "locally-produced" fractions contaminated
- **Subsistence Dairy Farmer**: Same as dairy farmer but fractions contaminated for fruits, vegetables, grains, and dairy is set to 1 rather than applying the "locally-produced" fractions contaminated

[a]Additionally, TRIM.Expo has the capability to assess infant exposures via the breast milk pathway for each/any of the listed scenarios.

TRIM.Expo$_{Ingestion}$ can incorporate measured concentrations (from monitoring data) as inputs. Because multimedia modeling involves many uncertainties, it may be helpful to implement a monitoring program to evaluate or further characterize key concentrations (e.g., those that drive the exposure estimate). Volume 1 (Chapter 19) provides an overview of multimedia modeling.

TRIM.Expo$_{Ingestion}$ provides outputs in the form of pollutant intake rate, oral intake rate, or Average Daily Dose (ADD) for each exposure scenario modeled. A future version of TRIM.Expo$_{Ingestion}$ will employ a population-based (vs. scenario-based) approach.

4.8.4   Risk Characterization

For this example Tier 3 approach, most risk estimates are calculated automatically by TRIM.Risk$_{HH}$, once the modeling scenarios are set up and run. The initial release of TRIM.Risk$_{HH}$ will calculate a deterministic ingestion cancer risk and non-cancer hazard metrics for the scenarios of interest.

This example Tier 3 approach considers background concentrations if these are a significant potential concern for a particular assessment. Again, note that consideration of background may or may not be appropriate pursuant to the specific legal and regulatory authorities under which the risk assessment is being conducted.

4.8.4.1   Reporting Results

The initial release of TRIM.Risk$_{HH}$ will provide a visualization tool for presenting analysis results in various automated formats. The risk assessor may want to utilize this tool to assist with development of the risk characterization summary, which generally will include the following information:

- Documentation of input parameters, outputs, and risk characterization, with special emphasis on the range of risk or hazard estimates;

- A simple presentation describing the assessment's purpose and the outcome relative to the purpose (e.g., purpose: demonstrate that risk is below target levels; outcome: low risk not demonstrated); and

- Documentation of all key assumptions or other inputs used for the assessment, such as emissions characteristics or choice of a nearby meteorological station. In particular, the rationale used to define personal profiles and how they are modeled and analyzed, need to be documented.

4.8.4.2   Assessment and Presentation of Uncertainty

Risk managers need to understand the strengths and the limitations of the Tier 3 multipathway risk assessment. A critical part of the risk characterization process, therefore, is an evaluation of the assumptions, limitations, and uncertainties inherent in the Tier 3 assessment in order to place the risk estimates in proper perspective.[3] Tier 3 multipathway risk assessments may be deterministic or probabilistic and commonly include semi-quantitative sensitivity analyses and quantitative uncertainty analysis (described in Section III-5.4 above). The general quantitative approach to propagating or tracking uncertainty through probabilistic modeling is described in Volume 1 (Chapter 31) of this reference library.

5.0   Ecological Risk Assessment

This section constitutes a snapshot of EPA's current thinking and approach to the adaptation of the evolving methods of ecological risk assessment to the context of Federal and state control of air toxics. While inhalation risk assessment has been increasingly used in regulatory contexts over the last several years, ecological risk assessment tools are less well developed and field tested in a regulatory context. This section should be considered a living document for review and input. By publishing this portion of Volume 2 in its current state of development, EPA is soliciting the involvement of persons with experience in this field to help improve these assessment methods for use in a regulatory context. EPA anticipates revisions to this draft section of Volume 2 on the basis of this input.

This section describes an example approach for performing facility/source-specific ecological risk assessments. In this example approach, ecological risk assessments are performed when air toxics that persist and which also may bioaccumulate or biomagnify in food chains (e.g., the PB-HAP compounds) are present in emissions; however, other factors such as state and local regulations may need to be considered in determining whether an ecological risk assessment is appropriate for a particular facility/source. Readers unfamiliar with ecological risk assessment are encouraged to consult the more general description of ecological risk assessment that is presented in Part IV of Volume 1.

5.1   Introduction and Overview

The ecological risk assessment process has three main phases that correspond to the three phases of the human health risk assessment methodology:[30]

*   **Planning, scoping, and problem formulation**, which focuses on identifying the ecological risk management goals, ecological receptors of concern, and assessment endpoints (explicit expression of the environmental value that is to be protected, operationally defined by an ecological entity and its attributes).

*   **Analysis** includes characterization of ecological effects for the PB-HAPs present in emissions from the facility/source. A distinction is made between assessment endpoints, which are the environmental values to be protected, and measures of effects, which are the specific metrics used to evaluate risk to the assessment endpoints. Common measures of effects include ecological toxicity reference values (Exhibit 33), which are discussed in more detail in Volume 1, Chapter 25. Analysis also includes characterization of direct exposures (e.g., estimated concentrations in abiotic media) or indirect exposures (based on dietary intake). Quantification of exposure via ingestion is similar to that for human health ingestion analyses, except that different food items may be involved, and the appropriate ecological exposure factors (e.g., diet, body weight) will be different.

    For this example approach, ecological effects analysis is limited to primary effects (e.g., lethality, reduced growth, neurological/behavioral and impaired reproduction) result from exposure of aquatic and terrestrial organisms to air toxics. An example of a chronic primary effect would be reduced reproduction in a fish species exposed to air toxics in a surface water body or mortality in a terrestrial bird eating contaminated fish from a small pond. Secondary effects (e.g., loss of prey species in the community) resulting from the action of air toxics on supporting components of the ecosystem are not included in this example approach. However, such effects could be examined in a Tier 3 analysis.

• Ecological **risk characterization** generally involves integration of exposure and stressor-response profiles with a summary of assumptions, scientific uncertainties, and strengths and limitations of the analyses. The final product is a risk description presenting the results of the integration, including an interpretation of ecological adversity and descriptions of lines of evidence. For the present purposes, risk characterization is often done more narrowly, comparing estimated media concentrations, dietary intake levels, or body burdens to ecological toxicity reference values using the hazard quotient or hazard index approach. Ecological risk characterization is described in more detail in Section 5.2 below.

Exhibit 33. Commonly Used Point Estimates

Median effect concentrations or doses (acute exposures)

LC_{50} Concentration (food or water) resulting in mortality in 50 percent of the exposed organisms
LD_{50} Dose (usually in dietary studies) resulting in mortality in 50 percent of the exposed organisms
EC_{50} Concentration resulting in a non-lethal effect (e.g., growth, reproduction) in 50 percent of the exposed organisms
ED_{50} Dose resulting in a non-lethal effect (e.g., growth, reproduction) in 50 percent of the exposed organisms

Low- or no-effect concentrations or doses (chronic exposures)

NOAEL   no-observed-adverse-effect-level, the highest dose for which effects are not statistically different from controls
LOAEL   lowest-observed-adverse-effect level, the lowest dose at which effects are statistically different from controls
NOEC   no-observed-effect-concentration, the highest ambient concentration for which adverse effects are not statistically different from controls
LOEC   lowest-observed-effect concentration, the lowest ambient concentration at which adverse effects are statistically different from controls
MATC   maximum acceptable toxicant concentration, the range of concentrations between the LOEC and NOEC
GMATC   geometric mean of the MATC, the geometric mean of the LOEC and NOEC

5.2   Ecological Risk Characterization

A common approach for characterizing ecological risks is the Hazard Quotient (HQ) approach (also referred to as the "quotient method"), which is similar to that used for human noncancer health risk assessment. In this approach, modeled or measured concentrations of the chemical in each environmental medium are divided by the appropriate ecological toxicity reference value (TRV) to yield a HQ for an individual chemical.

$$HQ = \frac{Oral\ Intake}{TRV} \quad or \quad HQ = \frac{EEC}{TRV} \quad or \quad HQ = \frac{BB}{TRV}$$

where:

HQ	=	hazard quotient.
Oral Intake	=	estimated or measured contaminant intake relevant to the oral intake-based ecological toxicity reference value (usually expressed as mg/kg-day).
TRV	=	ecological toxicity reference value. This may be in terms of oral intake, media concentration, or body burden. As described elsewhere, it may be a result of a single study (e.g., NOAEL) or the result of integration of multiple studies (e.g., water quality criterion).
EEC	=	estimated or measured environmental media concentration at the exposure point (usually expressed as mg/L for water and mg/kg for soil and sediment).
BB	=	estimated or measured body burden (usually expressed as mg/kg wet weight).

As with human health assessments, it is important that the measure of oral intake, EEC, or BB be in the same units as the ecological toxicity reference value to which the measure is being compared.

When ecological toxicity data for complex mixtures are unavailable, the hazard index (HI) approach[f] may be used, as scientifically appropriate, to integrate the ecological risks due to simultaneous exposure to multiple air toxics.

If the HI approach is used, the assumptions and associated limitations concerning air toxic interactions should be clearly documented. It may often be the case that a single chemical is responsible for the HI exceeding one, and the assessment can then focus on that chemical. In more refined assessments, the scientific integrity of assumptions inherent in the use of the HI will need to be carefully evaluated.

More complex approaches are available for characterizing ecological risks, as noted in Volume 1 (Part IV) of this reference library. If the risk assessment involves a regulatory decision, the risk assessor is encouraged to discuss any of these proposed approaches with the appropriate regulatory authorities prior to the analysis.

---

[f]The HI approach is termed the "quotient addition approach" in EPA's *Guidelines for Ecological Risk Assessment*[30-]

> ### Some Important Differences Between Ecological Risk Assessment and Multipathway Human Health Risk Assessment
>
> - **Planning and scoping.** The ecological risk assessment requires more preliminary analysis and deliberation regarding endpoints to be assessed and toxicity reference values to be used, because ecological systems are more complex and not as well understood biologically as are human health systems. The planning and scoping team should include individuals with specific expertise in ecological risk assessment.
>
> - **Assessment area.** It may be necessary to evaluate additional portions of the assessment area that are not of concern from a human health perspective.
>
> - **Potentially exposed populations.** The focus shifts from potentially exposed groups of humans to potentially exposed populations and species of ecological receptors of concern. In many cases, the exposure assessment may need to address multiple species and life-stages, many of which have physiological and biochemical processes that differ significantly from humans. (When threatened or endangered species are present, the assessment may also include an evaluation of those organisms as individuals).
>
> - **Exposure pathways and exposure routes.** It may be necessary to assess different exposure pathways and routes that are not of concern for human health.
>
> - **Ecological effects assessment.** Ecological systems have traits and properties that are different from humans and, thus, the ecological effects assessment (comparable to hazard assessment for human health) may consider a wider range of potential causal relationships.
>
> - **Risk characterization.** While risks may be assessed at multiple levels of ecological organization (i.e., organism, population, community, and ecosystem), they generally are assessed at the population level in air toxics assessments. (Nevertheless, when appropriate, consideration should be given to assessments as high levels of ecological organization, such as at the landscape level).

Ecological risk assessments are subject to additional sources of uncertainty and variability as compared to multipathway human health risk assessments. In addition to the uncertainties associated with multimedia modeling and sampling, the ecological risk assessment involves many decisions regarding choice of ecological receptors of concern and associated assessment endpoints and measures of effects. Some of these receptors may be at levels of organization above individual species (e.g., communities, ecosystems), where stressor-response relationships are poorly defined, characterized, and/or understood (e.g., air toxics effects on loss of community prey species mentioned in Section 5.1). Because many different species and higher taxonomic groups may be included in the assessment, selection of many parameter values such as bioconcentration factors, dose-response values, and dietary intake is more complex and uncertain for the ecological risk assessment as compared to the human health multipathway risk assessment.

5.3   Tier 1 Ecological Analysis

(reserved)

5.4   Tier 2 Ecological Analysis

5.4.1   Introduction

This section describes an example approach for performing a Tier 2 ecological risk assessment. Exhibit 34 provides an overview of this example approach. In this example approach, ISCST3 is used to provide deposition rates to soils and surface waters, and the MPE methodology is used to estimate PB-HAP concentrations in abiotic media (soils, surface waters, sediments) and wildlife food items (fish) at the most impacted location. These concentrations are compared to applicable ecological toxicity reference values to provide a conservative estimate of ecological risk to be calculated. If the facility/source passes this screening analysis, the risk manager can be reasonably confident that significant ecological risk is unlikely. As a consequence, monitoring is not included explicitly in this example Tier 2 ecological analysis (although available monitoring data can be used as inputs to the ecological exposure analysis)

5.4.2   Fate and Transport Modeling

This example approach uses the ISCST3 model for the Tier 2 fate and transport modeling.

5.4.2.1   Model Inputs

This example approach uses the same model inputs that were used for the example Tier 2 human health multipathway assessment.

5.4.2.2   Model Runs

This example approach uses the same model runs that were used for the example Tier 2 human health multipathway assessment.

Defining receptor locations

For this example Tier 2 analysis, terrestrial ecological receptors are assumed to occur at the modeling location ("node") with the maximum deposition rate as determined in the example Tier 2 human health multipathway assessment.

*   If multiple PB-HAPs are present, the point where maximum deposition occurs for each individual PB-HAP may not be in the same location. This example approach assumes that these are all co-located (see Section 5.4.4 below).

*   The point where maximum deposition occurs is used for terrestrial exposure pathways.

- For surface water pathways, this Tier 2 analysis is based on the surface water body within the assessment area where maximum concentrations occur. This was determined in the example Tier 2 human health multipathway analysis described earlier.

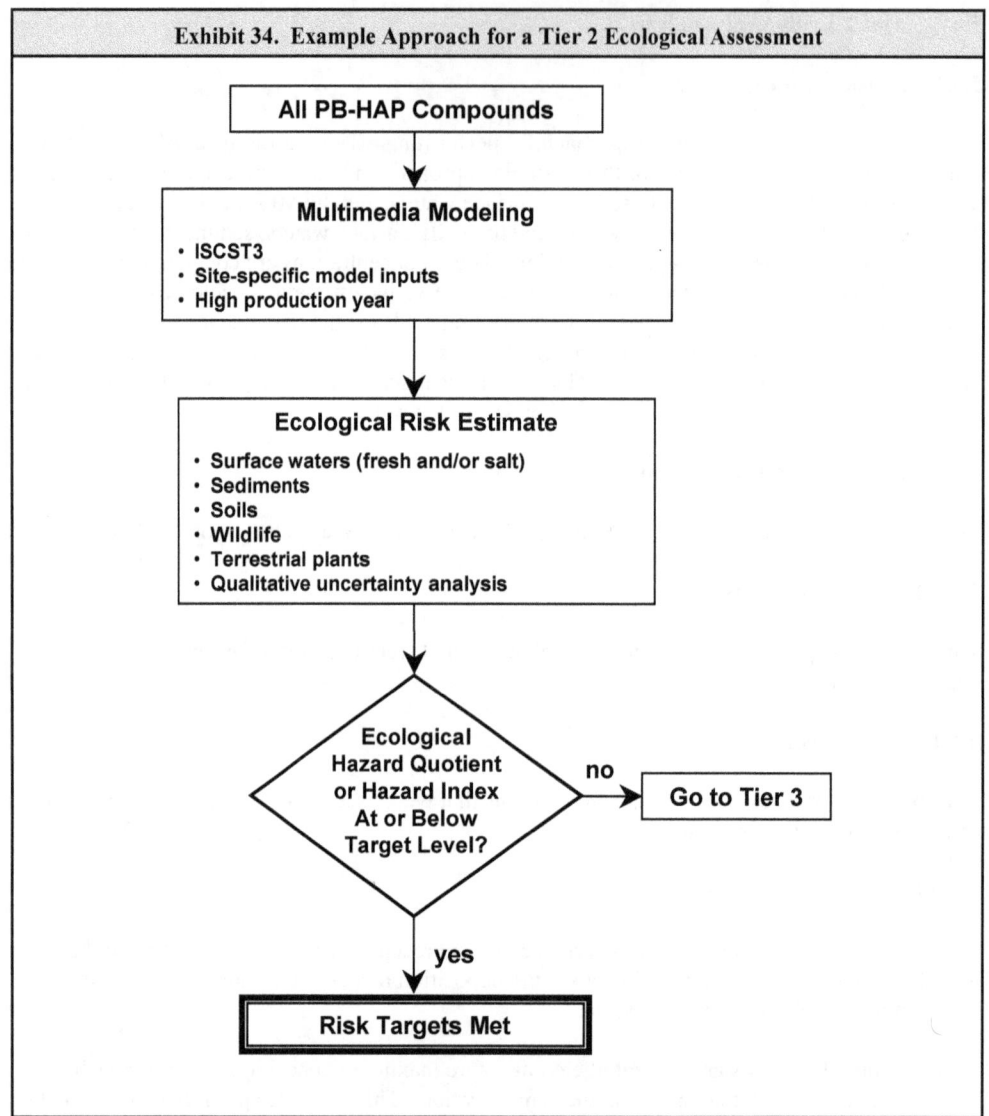

Exhibit 34. Example Approach for a Tier 2 Ecological Assessment

Defining the duration of emissions

This example approach uses the same duration of emissions that was used for the example Tier 2 human health multipathway analysis.

Defining deposition/flux rates for estimating media concentrations

This example approach uses the same deposition/flux rates that were used for the example Tier 2 human health multipathway analysis.

Model outputs to use

This example approach uses the same average annual deposition rates that were used for the example Tier 2 human health multipathway analysis as inputs to the MPE methodology.

5.4.3 Exposure Assessment

5.4.3.1 Characterization of Ecological Receptors

This example Tier 2 assessment uses generic ecological receptors. In other words, modeled concentrations of PB-HAPs in abiotic media and biota are compared to ecological toxicity reference values protective of the following groups of organisms: soil-dwelling biota, terrestrial plants, terrestrial animals, aquatic organisms, and sediment-dwelling biota. Note that EPA's ambient water quality criteria are protective of aquatic ecosystems (see Exhibit 35 below).

5.4.3.2 Defining the Point of Maximum Exposure

This example approach uses the same exposure locations used for the example Tier 2 human health multipathway analysis.

- For a single PB-HAP compound, the point of maximum exposure is defined as the modeled location (node) at which the highest deposition rate occurs (soils) and the surface water body with maximum modeled concentrations.

- For multiple PB-HAP compounds, the point of maximum exposure is defined by assuming that the point of maximum exposure occurs at the same location for each PB-HAP compound (i.e., exposure for each compound is added to calculate total exposure).

5.4.3.3 Calculation of Exposure Concentration

This example approach uses the following concentrations estimated for the example Tier 2 multipathway human health assessment:

- Maximum concentrations in soil, surface water, and sediment; and
- Maximum concentration in fish.

Note that existing monitoring data may be used to evaluate or refine the exposure estimates based on multimedia modeling. Volume 1 provides an overview of multimedia monitoring (Chapter 19) and additional discussion of monitoring for ecological exposure analysis (Chapter 24).

5.4.3.4   Determining Intake

Ecological toxicity reference values for some PB-HAPs incorporate assumptions about wildlife intake. For those substances, no additional calculations are required to determine intake. For the remaining PB-HAPs, it will be necessary to determine intake. Equations and exposure factors used to calculate intake are found in EPA's *Wildlife Exposure Factors Handbook*.[31]

5.4.4   Risk Characterization

This example approach considers three types of comparisons:

- Comparison of modeled concentrations in soils, surface waters, and sediments to ecological toxicity reference values for each medium;

- Comparison of dietary intake levels for terrestrial birds and mammals to ecological toxicity reference values for wildlife ingestion; and

- Comparison of estimated body burdens in fish to ecological toxicity reference values that relate body burdens to adverse ecological impacts.

For these comparisons, three general types of ecological toxicity reference values are applicable (Exhibit 35):

- Toxicity reference values that relate concentrations of PB-HAPs in abiotic media (e.g., soil, sediment) to adverse effects (these may be available at the species- or community-level);

- Toxicity reference values that relate dietary intake levels of PB-HAPs to adverse effects (e.g., in birds and mammals); and

- Toxicity reference values that relate concentrations of PB-HAPs in biota (i.e., body burdens) to adverse effects in those biota. These generally have to be derived based on oral toxicity data and assumptions about wildlife exposure factors.

Exhibit 35. Sources of Ecological Toxicity Reference Values (TRVs) or Benchmarks

Data Source	Freshwater	Saltwater	Sediment	Soil/ Terrestrial	Reference
TRVs or benchmarks based on ambient concentrations					
EPA Ambient Water Quality Criteria	X	X			EPA has developed national recommended water quality criteria for the protection of aquatic life for approximately 150 pollutants. These criteria are published pursuant to Section 304(a) of the Clean Water Act (CWA) and provide guidance for States and Tribes to use in adopting water quality standards under Section 303(c) of the CWA. Source: http://www.epa.gov/waterscience/criteria/aqlife.html
Sediment Quality Criteria			X		EPA and other agencies have developed sediment quality criteria for the protection of benthic communities. These criteria are highly specific to regions and bodies of water in the U.S. Regional experts are the recommended source for appropriate facility/source-specific criteria.
NOAA Screening Quick Reference Tables (SQuiRTs)	X	X	X	X	NOAA has developed a set of Screening Quick Reference Tables, or SQuiRTs, that present screening concentrations for inorganic and organic contaminants in soil, sediment, and surface water. The SQuiRTs also include guidelines for preserving samples and analytical technique options. Note that sediment SQuiRTs may not be appropriate to facility/source-specific bodies of water; consultation with regional experts is recommended. http://response.restoration.noaa.gov/cpr/sediment/squirt/squirt.html.

Exhibit 35. Sources of Ecological Toxicity Reference Values (TRVs) or Benchmarks

Data Source	Freshwater	Saltwater	Sediment	Soil/ Terrestrial	Reference
Great Lakes Criteria	X				The GLWQI Tier II criteria and SCV have received some peer review prior to publication, and 12 of them are included in the HWIR, which underwent public comment before promulgation. The GLWQI Tier II methodology calculates SCV in a similar way to FCV, but uses statistically derived "adjustment factors" and has less rigorous data requirements. • Tier II Criteria are designed to be protective of aquatic communities • SCV are designed to measure chronic toxicity to aquatic organisms Source: *Ecotox Thresholds ECO Update* (volume 3, No. 2, January 1996, EPA/540/F-95/038).
EPA Soil Screening Levels				X	EPA has developed a methodology and initial soil screening levels protective of ecological receptors. Source: United States Environmental Protection Agency. 2000. *Ecological Soil Screening Guidance (Draft).* Office of Emergency and Remedial Response, Washington, D.C., July 2000. http://www.epa.gov/superfund/programs/risk/ecorisk/ecossl.htm.
EPA Region 4 Soil Screening Levels				X	Source: U.S. Environmental Protection Agency 1995. Supplemental Guidance to RAGS: Region 4 Bulletins No. 2. Ecological Risk Assessment. Region IV, Waste Management Division. http://www.epa.gov/region04/waste/ots/ecolbul.htm

Data Source	Freshwater	Saltwater	Sediment	Soil/ Terrestrial	Reference
Netherlands target values				X	The Netherlands bases the prevention and remediation of contaminated soil on its Soil Protection Act. Target, limit, and intervention values have been established for soil and groundwater as part of a general framework of risk-based environmental quality objectives. Target values represent background concentrations in which risk is considered negligible. If target values are currently not met, limit values may be applied to define general concentrations which must be attained. Source: http://www.contaminatedland.co.uk/std-guid/dutch-l.htm

TRVs or benchmarks based on body burdens (tissue concentrations)

USACE/ EPA ERED	X	X			The U.S. Army Corps of Engineers/U.S. Environmental Protection Agency Environmental Residue-Effects Database (ERED) is a compilation of data, taken from the literature, where biological effects (e.g., reduced survival, growth, etc.) and tissue contaminant concentrations were simultaneously measured in the same organism. Currently, the database is limited to those instances where biological effects observed in an organism are linked to a specific contaminant within its tissues. Source: http://www.wes.army.mil/el/ered/index.html

Exhibit 35. Sources of Ecological Toxicity Reference Values (TRVs) or Benchmarks

Data Source	Freshwater	Saltwater	Sediment	Soil/Terrestrial	Reference
Toxicity data that can be used to derive TRVs or benchmarks for dietary intake					
ECOTOX	X	X	X	X	ECOTOX is a source for locating single chemical toxicity data for aquatic life, terrestrial plants and wildlife. ECOTOX was created and is maintained by EPA's Office of Research and Development and the National Health and Environmental Effects Research Laboratory's Mid-Continent Ecology Division. ECOTOX is a source for locating single chemical toxicity data from three EPA ecological effects databases: AQUIRE, TERRETOX, and PHYTOTOX. AQUIRE and TERRETOX contain information on lethal, sublethal and residue effects. AQUIRE includes toxic effects data on all aquatic species including plants and animals and freshwater and saltwater species. TERRETOX is the terrestrial animal database. It primarily focuses on wildlife species but the database does include information on domestic species. PHYTOTOX is a terrestrial plant database that includes lethal and sublethal toxic effects data. Source: http://www.epa.gov/ecotox.
CAL-Ecotox	X	X	X	X	The California Wildlife Biology, Exposure Factor, and Toxicity Database (Cal/Ecotox) is a compilation of ecological, physiological data, and toxicity data for a number of California mammals, birds, amphibians and reptiles. http://www.oehha.ca.gov/cal_ecotox

Exhibit 35. Sources of Ecological Toxicity Reference Values (TRVs) or Benchmarks

Data Source	Freshwater	Saltwater	Sediment	Soil/ Terrestrial	Reference
ORNL Toxicity databases				X	Oak Ridge National Laboratory has developed toxicity reference values for soils protective of terrestrial plants, soil and litter invertebrates, and terrestrial wildlife. Sources: Efroymson, R.A., M.E. Will, G.W. Suter II, and A.C. Wooten. 1997. *Toxicological Benchmarks for Screening Contaminants of Potential Concern for Effects on Terrestrial Plants: 1997 Revision.* Prepared for the U.S. Department of Energy, Office of Environmental Management. Oak Ridge National Laboratory, Oak Ridge, TN. ES/ER/TM-85/R3. http://www.hsrd.ornl.gov/ecorisk/tm85r3.pdf Efroymson, R.A., M.E. Will, and G.W. Suter II. 1997. *Toxicological Benchmarks for Contaminants of Potential Concern for Effects on Soil and Litter Invertebrates and Heterotrophic Process: 1997 Revision.* Prepared for the U.S. Department of Energy, Office of Environmental Management. Oak Ridge National Laboratory, Oak Ridge, TN. ES/ER/TM-126/R2. http://www.hsrd.ornl.gov/ecorisk/tm126r21.pdf Efroymson, R.A., G.W. Suter II, B.E. Sample, and D.S. Jones 1997. *Preliminary Remediation Goals for Ecological Endpoints.* Prepared for the U.S. Department of Energy, Office of Environmental Management. Oak Ridge National Laboratory, Oak Ridge, TN. ES/ER/TM-162/R2. http://www.hsrd.ornl.gov/ecorisk/tm162r2.pdf

Exhibit 35. Sources of Ecological Toxicity Reference Values (TRVs) or Benchmarks

Data Source	Freshwater	Saltwater	Sediment	Soil/ Terrestrial	Reference
ECOSAR	X	X			ECOSAR is a computer program that uses structure-activity relationships (based on available data) to predict the acute and chronic toxicity of organic chemicals to aquatic organisms. ECOSAR provides quantitative estimates of chronic values (e.g., GMATC), acute LC_{50} values, and acute EC_{50} values for industrial chemicals for several aquatic species (e.g., fish, daphnia, green algae, mysids). When the estimated aquatic toxicity reference value exceeds the water solubility of the compound, the estimated value is flagged; this situation generally is interpreted to mean that the chemical has no toxic effects in a saturated solution. Source: http://www.epa.gov/oppt/newchems/21ecosar.htm

5.4.4.1   Reporting Results

For this example approach, a relative simple summary can be used to report results, as long as it is consistent with the need to make the results both transparent and reproducible. The summary generally will include the following information:

- Documentation of input parameters, output values, and risk characterization, with special emphasis on comparing estimated risk/hazard to risk targets;

- A simple presentation describing the assessment's purpose (e.g., to determine whether risk is below levels of concern) and the outcome relative to that purpose (e.g., low risk is not demonstrated); and

- Documentation of anything that is discretionary (i.e., anything that is facility-specific), such as emissions characteristics or choice of a meteorological station other than the nearest.

5.4.4.2   Assessment and Presentation of Uncertainty

Risk managers need to understand the strengths and the limitations of the Tier 2 ecological risk assessment. A critical part of the risk characterization process, therefore, is an evaluation of the assumptions, limitations, and uncertainties inherent in the Tier 2 assessment in order to place the risk estimates in proper perspective.[3] Tier 2 ecological assessments commonly include a quantitative or qualitative description of the uncertainty for each parameter and indicating the possible influence of these uncertainties on the final risk estimates given knowledge of the models used. Tier 2 assessments also may include a semi-quantitative sensitivity analyses. These approaches are described in Section III-5.4 above. Sensitivity analyses are discussed in more detail in Volume 1 (Chapters 3 and 13) of this reference library.

5.5   Tier 3 Ecological Analysis

5.5.1   Introduction

This section describes an example approach for performing a Tier 3 ecological risk assessment. Exhibit 36 provides an overview of this example approach. This example Tier 3 assessment is significantly different than the Tier 2 example approach, in that it involves the use of TRIM.FaTE for multimedia modeling, specific consideration of ecological receptor locations, and use of a risk model (TRIM.Risk$_{Eco}$) to provide a more refined characterization of risk, including multiple estimates of risk.

This example Tier 3 analysis allows considerable flexibility in analytical approach and detail.

- In this example approach, TRIM.FaTE is used for multimedia modeling and generates concentrations in abiotic media, body burdens, and wildlife intake rates that feed directly into TRIM.Risk$_{Eco}$ for ecological risk characterization. TRIM.FaTE allows considerable spatial refinement in selecting exposure locations, and provides concentrations for user-specified (actual) locations.

- In this example approach, a risk model (TRIM.Risk$_{Eco}$) is used to characterize ecological risk. This module performs ecological risk characterization calculations for multimedia ecological risk assessments of toxic air pollutants or other chemical contaminants. The output of TRIM.Risk$_{Eco}$ are hazard quotients (HQ) that can be processed with TRIM.Risk analysis tools to prepare tables and charts that can be used in communicating the risk assessment results.

- In this example approach, monitoring data are used to evaluate or further characterize key concentration data.

Exhibit 37 illustrates how the TRIM modules and data correspond to EPA's ecological risk assessment framework as described in the *Framework for Ecological Risk Assessment*.[32] Guidance for how to structure TRIM analyses for ecological risk assessment is provided in the *TRIM.Risk$_{Eco}$ User's Guide*.[33] This example Tier 3 analysis will be highly facility/source-specific and will require careful planning.

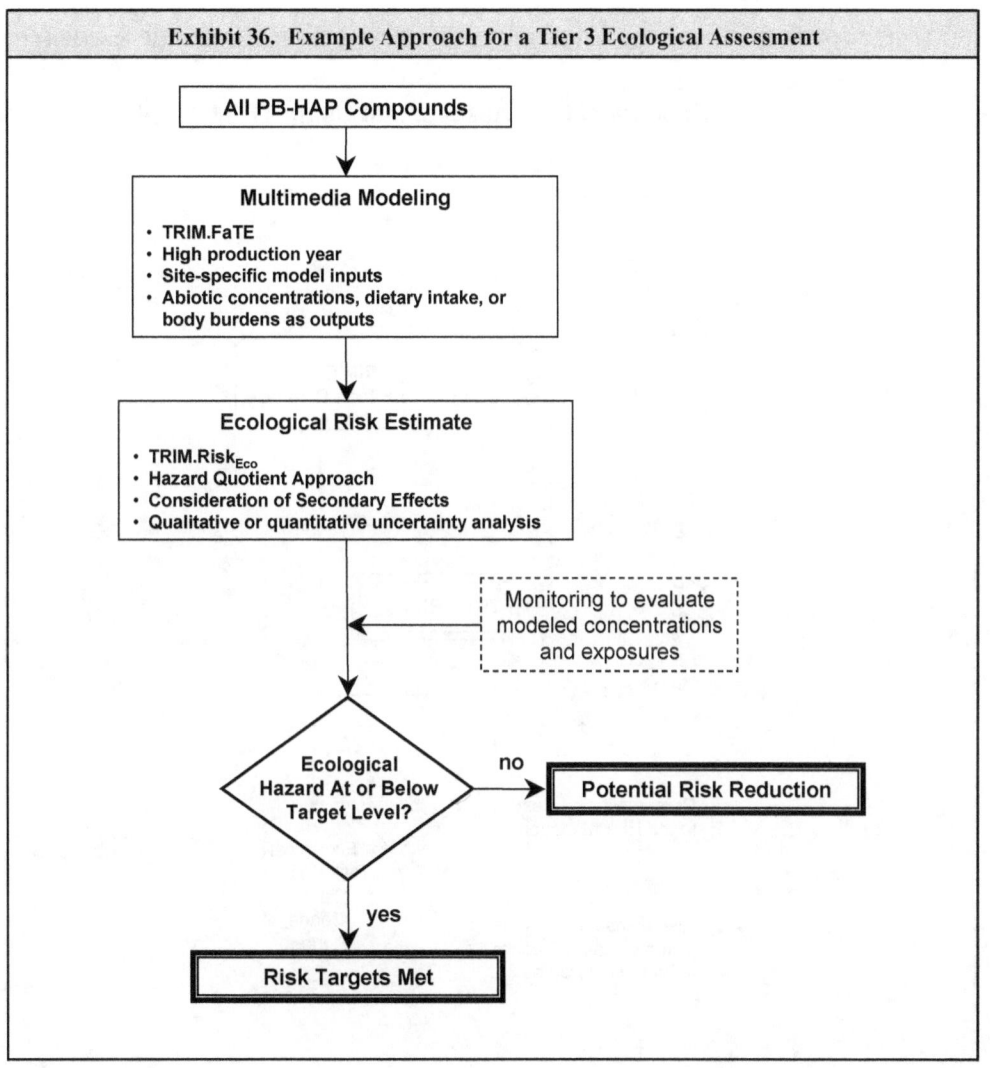

Exhibit 36. Example Approach for a Tier 3 Ecological Assessment

All PB-HAP Compounds

Multimedia Modeling
- TRIM.FaTE
- High production year
- Site-specific model inputs
- Abiotic concentrations, dietary intake, or body burdens as outputs

Ecological Risk Estimate
- TRIM.Risk$_{Eco}$
- Hazard Quotient Approach
- Consideration of Secondary Effects
- Qualitative or quantitative uncertainty analysis

Monitoring to evaluate modeled concentrations and exposures

Ecological Hazard At or Below Target Level?

no → **Potential Risk Reduction**

yes → **Risk Targets Met**

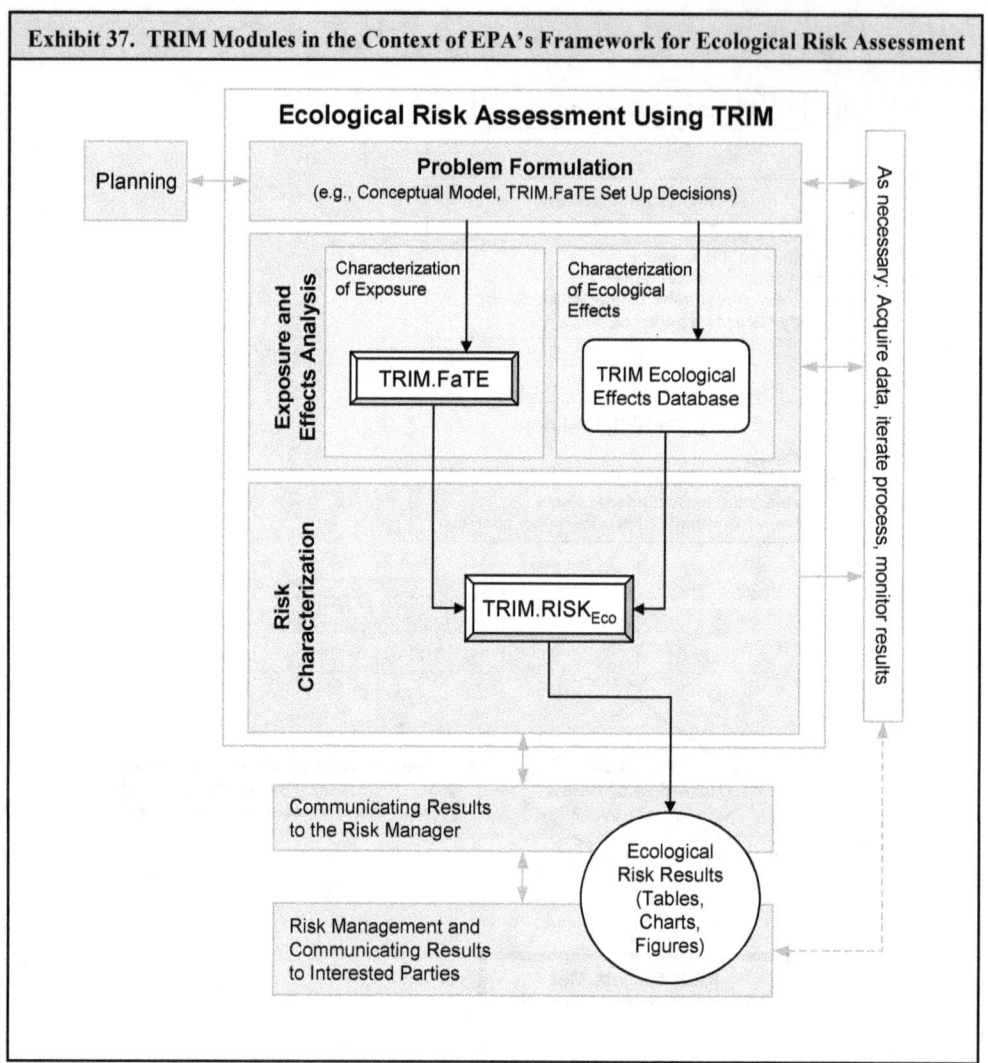

5.5.2 Identification of Potentially Exposed Populations

For this example approach, potentially exposed populations (ecological receptors of concern) may be defined at the species and/or community level (Exhibit 38). Within these general categories, users have considerable flexibility in defining ecological receptors. For example, the analysis could focus on a particular species (e.g., a mink), or the species could be used as a surrogate for an entire trophic group (e.g, terrestrial carnivores). Specific receptors may be of concern for a variety of reasons, including:

- The receptor (or one of it's life stages) is particularly vulnerable to or sensitive to one or more PB-HAP compounds;

- The receptor (usually a species or a community such as a wetland) is listed as endangered or threatened or is otherwise given special legal protection by the state or federal government;

- The receptor plays an important part in the overall structure or function of the ecological community or ecosystem; and/or

- The receptor is of particular economic or cultural value to local populations.

Exhibit 38. Potential Ecological Receptors of Concern	

Species-level	Terrestrial plants
	Terrestrial invertebrates
	   • soil-dwelling (e.g., earthworms)
	   • surface-dwelling
	Terrestrial vertebrates
	   • birds
	   • mammals
	   • reptiles
	   • amphibians
	Aquatic plants
	Aquatic invertebrates
	   • sediment-dwelling
	   • benthic
	   • pelagic
	Aquatic vertebrates
	   • fish
	   • (reptiles)
	   • (amphibians)

Community-level	Aquatic communities Wetland communities

Often it is important to understand the aquatic and terrestrial food webs in the habitats of concern, because these can be important parts of ecological exposure pathways. Top predators are often of special concern for exposure to PB-HAP compounds.

For this example approach, ecological receptors for each habitat potentially impacted are identified to ensure (1) plant and animal communities representative of the habitat are represented by the habitat-specific food web, and (2) potentially complete exposure pathways are identified. Ecological receptor identification may need to include species both known and expected to be present in a specific habitat being evaluated, and include resident and migratory populations. Consultation with ecological experts is encouraged (see Volume 1, Chapter 23).

5.5.3   Assessment Endpoints and Measures of Effect

As discussed in Volume 1 of this reference library (Chapter 23), an **assessment endpoint** is an explicit expression of the "actual environmental value that is to be protected" or is of concern. It includes the identification of the ecological entity for the analysis (e.g., a species, ecological resource, habitat type, or community) as well as the attribute of that entity that is potentially at risk (e.g., reproductive success, production per unit area, surface area coverage, biodiversity) and that is important to protect. **Measures of effects** are the metrics by which these endpoints are assessed.[30] EPA has recently released guidance that describes a set of endpoints, known as Generic Ecological Assessment Endpoints (GEAE), that can be considered and adapted for specific ecological risk assessments (Exhibit 39).[34]

Where ecological effects information is available, TRIM.Risk$_{Eco}$ allows the user to select (or input) multiple measures of effects for a given ecological receptor. For example, different toxicity reference values (TRVs) may be available for the same species that are associated with different measures of effects (e.g., reproductive effects, respiration, mortality). Volume 1 (Chapter 25) provides a discussion on the selection of TRVs for a particular assessment.

Where available ecological toxicity reference values do not consider secondary effects (e.g., on communities), it may be possible to use TRIM.Risk$_{Eco}$ outputs as a starting point for evaluating secondary effects, either qualitatively or quantitatively. For this example Tier 3 assessment, risk assessors are encouraged to include secondary effects as assessment endpoints to the extent that resources and information allows.

Exhibit 39. Generic Ecological Assessment Endpoints[a]		
Entity	**Attribute**	**Identified EPA Precedents**
Organism-level endpoints		
Organisms (in an assessment population or community)	Kills (mass mortality, conspicuous mortality)	Vertebrates
	Gross anomalies	Vertebrates, shellfish, plants
	Survival, fecundity, growth	**Endangered species, migratory birds, marine mammals, bald and golden eagles**, vertebrates, invertebrates, plants
Population-level endpoints		
Assessment population	Extirpation	Vertebrates
	Abundance	Vertebrates, shellfish
	Production	Vertebrates (game/resource species), harvested plants
Community and ecosystem-level endpoints		
Assessment communities, assemblages, and ecosystems	Taxa richness	Aquatic communities, **coral reefs**
	Abundance	Aquatic communities
	Production	Plant assemblages
	Area	**Wetlands, coral reefs**, endangered/rare ecosystems
	Function	**Wetlands**
	Physical structure	Aquatic ecosystems
Officially designated endpoints		
Critical habitat for endangered or threatened species	Area Quality	
Special places	Ecological properties that relate to the special or legally protected properties	e.g., **National Parks, National Wildlife Refuges, Great Lakes**
[a]Generic ecological assessment endpoints for which EPA has identified existing policies and precedents (in particular, the specific entities listed in the third column). Bold indicates protection by federal statute. *Source*: EPA's *Generic Ecological Assessment Endpoints (GEAE) for Ecological Risk Assessment*[34]		

5.5.4 Fate and Transport Modeling

This subsection provides an overview of TRIM.FaTE set up and execution. This is explained more fully in section 4.8.2 and in the *TRIM.FaTE User's Guide*.[21]

- The **modeling region** encompasses all areas where ecological receptors are known or likely to occur, especially any important receptors such as wetlands, endangered/threatened species, or areas that are particularly important to regional populations (e.g., wintering areas, migration routes, breeding grounds).

- **Parcels** are defined to adequately provide an assessment of exposures at the locations where the ecological receptors identified above occur.

- **Volume elements** are defined to include all potential routes of ecological exposure (e.g., root zone soil for earthworms).

- **Compartments** can be defined in a flexible manner depending on the level of detail in which ecological receptors are characterized for the analysis. For example, a single species (e.g., lake trout) could be used as a surrogate for all higher-trophic level fish, or the analysis could identify several multiple species of higher-trophic level fish.

- **Links and algorithms** should incorporate the processes that drive chemical transfer and transformation that are relevant to the ecological exposure assessment.

- As noted above, the TRIM.FaTE model runs may be set up and performed along with the Tier 3 human health model runs.

- Results can be processed with TRIM.Risk analysis or other tools to prepare tables and charts that can be used in communicating the risk assessment results.

5.5.5 Exposure Assessment

TRIM.FaTE provides three general metrics of ecological exposure (Exhibit 40): ambient concentrations in abotic media, ingestion intake levels for specified receptors, and body burdens. Each of these can be compared with a corresponding ecological toxicity reference value for risk characterization. Intake and body burdens are calculated for specific ecological receptors specified in the model setup, although these receptors may be matched with TRVs for surrogates, as appropriate, in TRIM.Risk$_{Eco}$.

The user can specify the specific **time points of interest** of the TRIM.FaTE exposure simulation to be compared with each type of ecological toxicity reference values.

Exhibit 40. TRIM Metrics of Ecological Exposure and Metrics of Ecological Effects	
Metrics of Ecological Exposure Available from TRIM.FaTE	**Corresponding Ecological Toxicity reference value Available in the TRIM Ecological Effects Database**
Ambient concentrations in abiotic compartments of interest (i.e., soil, water, air, and sediment concentrations)	Abiotic media concentrations associated with adverse effects
Ingestion intake (on a mass per mass and time basis) for ecological receptor(s) of interest	Biota oral intake (on a mass per mass and time basis) associated with adverse effects
Whole-body concentrations in biotic compartments of interest (i.e., body burdens)	Body burden concentrations associated with adverse effects

Because ecological exposure assessments are subject to many sources of uncertainty and variability (e.g., uncertainties associated with multimedia modeling; choice of ecological receptors of concern and associated assessment endpoints and measures of effects), it may be helpful to incorporate a multimedia monitoring program to evaluate or refine the exposure estimates based on multimedia modeling. Volume 1 provides an overview of multimedia monitoring (Chapter 19) and additional discussion of monitoring for ecological exposure analysis (Chapter 24).

5.5.6 Risk Characterization

TRIM.Risk$_{Eco}$ quantifies the potential for ecological risk from exposure to PB-HAPs using a hazard quotient (HQ) approach. The HQs calculated by TRIM.Risk$_{Eco}$ are based on metrics of ecological exposure calculated in TRIM.FaTE and the corresponding metrics of ecological effects compiled in the TRIM Ecological Effects database (Exhibit 41). TRIM.Risk$_{Eco}$ can also estimate HQs based on ecological exposure estimates from models other than TRIM.FaTE, assuming the data are provided in the TRIM.FaTE MySQL database format. Detailed instructions on configuring this module are included in the *TRIM.Risk$_{Eco}$ User's Guide*.[33]

Users can also calculate HQs for other ecological receptors whose TRVs are based on **abiotic media concentrations**. For instance, the ecological TRVs database may include a TRV for largemouth bass that is based on the pollutant concentration in surface water. If a user modeled surface water concentrations in TRIM.FaTE, s/he could estimate the HQ for largemouth bass for this chemical even if large mouth bass were not included in the TRIM.FaTE simulation. This feature allows users to calculate HQs for a variety of receptors without including these receptors in exposure modeling.

The TRIM Ecological Effects database contains a variety of ecological toxicity reference values for PB-HAPs. Volume 1 of this reference library (Chapter 26) provides a listing of additional sources of ecological toxicity reference value levels.

5.5.6.1　Reporting Results

TRIM.Risk provides a visualization tool for presenting analysis results in various automated formats. The risk assessor may want to utilize this tool to assist with development of the risk characterization summary, which generally will include the following information:

- Documentation of input parameters, outputs, and risk characterization, with special emphasis on the range of risk or hazard estimates;

- A simple presentation describing the assessment's purpose and the outcome relative to the purpose (e.g., purpose: demonstrate that risk is below target levels; outcome: low risk not demonstrated); and

- Documentation of all key assumptions or other inputs used for the assessment, such as emissions characteristics or choice of a nearby meteorological station.

5.5.6.2　Assessment and Presentation of Uncertainty

Risk managers need to understand the strengths and the limitations of the Tier 3 ecological risk assessment. A critical part of the risk characterization process, therefore, is an evaluation of the assumptions, limitations, and uncertainties inherent in the Tier 3 assessment in order to place the risk estimates in proper perspective.[3] Tier 3 ecological risk assessments may be deterministic or probabilistic and commonly include semi-quantitative sensitivity analyses and quantitative uncertainty analysis (described in Section III-5.4 above). The general quantitative approach to propagating or tracking uncertainty through probabilistic modeling is described in Volume 1 (Chapter 31) of this reference library.

References (note that some numbered references include multiple citations)

1. U.S. Environmental Protection Agency. 2003. *Framework for Cumulative Risk Assessment.* Risk Assessment Forum, Washington, DC 20460, May 2003, EPA/630/P-02/001F; available at http://cfpub.epa.gov/ncea/raf/recordisplay.cfm?deid=54944.

2. U.S. Environmental Protection Agency. 1999. *Residual Risk Report to Congress.* Office of Air Quality Planning and Standards, Research Triangle Park, NC 27711, March 1999, EPA-453/R-99-001; available at http://www.epa.gov/ttn/oarpg/t3/reports/risk_rep.pdf.

3. U.S. Environmental Protection Agency. 1995. *Policy for Risk Characterization at the U.S. Environmental Protection Agency.* Science Policy Council, Washington, D.C., March 1995, available at: http://64.2.134.196/committees/aqph/rcpolicy.pdf.

4. Hattis. D.B. and D.E. Burmaster. 1994. Assessment of variability and uncertainty distributions for practical risk analyses. *Risk Analysis* 14(5):713-730.

   U.S. Environmental Protection Agency. 1999. *Report of the Workshop on Selecting Input Distributions for Probabilistic Assessments.* Risk Assessment Forum, Washington, DC. EPA/630/R-98/004.

   U.S. Environmental Protection Agency. 1997. *Guiding Principles for Monte Carlo Analysis.* Risk Assessment Forum, Washington, DC. March 1997. EPA-630/R-97-001; available at http://www.epa.gov/NCEA/pdfs/montcarl.pdf.

   U.S. Environmental Protection Agency. 1988. *National Emission Standards for Hazardous Air Pollutants. Federal Register* 53(145):28496-28056, Proposed Rule and Notice of Public Hearing, July 28, 1988.

   Finkel, A.M. 1990. *Confronting Uncertainty in Risk Management: A Guide for Decision-Makers.* Center for Risk Management, Resources for the Future. Washington, DC.

   U.S. Environmental Protection Agency. 1997. *Policy for Use of Probabilistic Analysis in Risk Assessment.* Office of the Administrator, Washington, DC. May 15, 1997; available at http://www.epa.gov/osp/spc/probpol.htm.

   National Council on Radiation Protection and Measurements. 1996. *A Guide for Uncertainty Analysis in Dose and Risk Assessments Related to Environmental Contamination.* NCRP Commentary No. 14; available at http://www.ncrp.com/comm14.html.

5. U.S. Environmental Protection Agency. 2001. *Risk Assessment Guidance for Superfund (RAGS). Volume III - Part A. Process for Conducting Probabilistic Risk Assessment.* Office of Emergency and Remedial Response, Washington, D.C. (http://www.epa.gov/superfund/programs/risk/rags3a/index.htm).

6. U.S. Environmental Protection Agency. 1992. *Screening Procedures for Estimating the Air Quality Impact of Stationary Sources, Revised.* Office of Air Quality Planning and Standards, Research Triangle Park, North Carolina, October 1992. EPA-454/R-92-019. Available at: http://www.epa.gov/scram001/tt25.htm#guidance.

7. U.S. Environmental Protection Agency. 2002. *User's Guide for the Human Exposure Model (HEM).* Office of Air Quality Planning and Standards, Research Triangle Park, NC; current version available at: http://www.epa.gov/ttn/fera/human_hem.html.

8. U.S. Environmental Protection Agency. 1995. *SCREEN3 Model Users' Guide.* Office of Air Quality Planning and Standards, Research Triangle Park, NC. EPA-454/B-95-004. (http://www.epa.gov/scram001/userg/screen/screen3d.pdf).

9. U.S. Environmental Protection Agency. 1995. *User's Guide for the Industrial Source Complex (ISC3) Dispersion Models Volume I - User's Instructions.* Office of Air Quality Planning and Standards, Research Triangle Park, NC. September 1995. EPA-454/B-95-0031; available at http://www.epa.gov/scram001/tt22.htm#isc.

10. U.S. Environmental Protection Agency. 2003. *Total Risk Integrated Methodology Trim.Expo$_{Inhalation}$ User's Document. Volume 1: Air Pollutants Exposure Model (APEX, Version 3) User's Guide.* Office of Air Quality Planning and Standards, Research Triangle Park, NC. April 2003.

11. EPA's carcinogen risk assessment guidelines were first published in 1986, revisions were proposed in 1996 and 2001, the July 1999 draft of the revisions were adopted as interim guidance. A subsequent 2000 draft of the Guidelines has been released for public and scientific review prior to adoption as final (see http://cfpub.epa.gov/ncea/raf/rafguid.htm).

   U.S. Environmental Protection Agency. 1999. *Guidelines for Carcinogen Risk Assessment. Review Draft.* Risk Assessment Forum, Washington, DC. NCEA-F-0644.

   U.S. Environmental Protection Agency. 1986. *Guidelines for Carcinogen Risk Assessment.* Federal Register 51(185):33992-43003.

   U.S. Environmental Protection Agency. 2003. *Draft Final Guidelines for Carcinogen Risk Assessment (External Review Draft)*, Risk Assessment Forum, Washington, DC NCEA-F-0644A.

12. U.S. Environmental Protection Agency. *Health Effects Information Used in Cancer and Noncancer Risk Characterization for the NATA 1996 National-Scale Assessment*, available at http://www.epa.gov/ttn/atw/nata/nettables.pdf.

13. U.S. Environmental Protection Agency. 1998. *An SAB Report: Development of the Acute Reference Exposure.* Science Advisory Board, Washington, D.C., November 1998, available at: http://www.epa.gov/sab/pdf/ehc9905.pdf.

14. U.S. Environmental Protection Agency. 2000. *Supplementary Guidance for Conducting Health Risk Assessment of Chemical Mixtures*. Risk Assessment Forum, Washington, D.C. EPA/630/R-00/002, available at http://www.epa.gov/ncea/raf/pdfs/chem_mix/chem_mix_08_2001.pdf.

U.S. Environmental Protection Agency. 1986. *Guidelines for the Health Risk Assessment of Chemical Mixtures*. EPA/630/R-98/002; published in the Federal Register 51 (185):34014-34025, Sept 24, 1986, available at http://www.epa.gov/ncea/raf/pdfs/chem_mix/chem_mix_08_2001.pdf.

15. Morgan, G. and Henrrion. M. 1990. *Uncertainty: A Guide to Dealing with Uncertainty in Quantitative Risk and Policy Analysis*. Cambridge University Press, Cambridge, U.K.

16. U.S. Environmental Protection Agency. 1989. *Risk Assessment Guidance for Superfund: Volume I. Human Health Evaluation Manual (Part A)*. Office of Emergency and Remedial Response. Washington, D.C., EPA/541/1-89/002, available at: http://www.epa.gov/superfund/programs/risk/ragsa/index.htm

17. U.S. Environmental Protection Agency. 2003. *Guideline on Air Quality Models*. 40 CFR Part 51, Appendix W, available at http://www.epa.gov/scram001/tt25.htm#guidance.

See also *Procedures for Air Modeling of Superfund Sites* and *Dispersion Modeling of Toxic Pollutants in Urban Areas*, available at (http://www.epa.gov/ttn/scram/).

18. The CHAD database is available at http://www.epa.gov/chadnet1/.

19. U.S. Environmental Protection Agency. 1998. *Methodology for Assessing Health Risks Associated with Multiple Pathways of Exposure to Combustor Emissions*. Update to EPA/600/6-90/003 *Methodology for Assessing Health Risks Associated With Indirect Exposure to Combustor Emissions*. National Center for Environmental Assessment. EPA-600/R-98-137, available at: http://cfpub.epa.gov/ncea/cfm/recordisplay.cfm?deid=55525.

20. U.S. Environmental Protection Agency. 1998. *Human Health Risk Assessment Protocol for Hazardous Waste Combustion Facilities, Peer Review Draft*. Office of Solid Waste and Emergency Response, July 1998. EPA530-D-98-001 A, B, and C and subsequent Errata (EPA Memo, July 1999), available at http://www.epa.gov/earth1r6/6pd/rcra_c/pd-o/comb_risk.htm.

21. U.S. Environmental Protection Agency. 2002. *TRIM.FaTE Technical Support Document, Volume 1: Description of Module*. Office of Air Quality Planning and Standards, Research Triangle Park, NC. EPA 453/R-02-011a, available at http://www.epa.gov/ttn/fera/.

U.S. Environmental Protection Agency. 2002. *TRIM.FaTE Technical Support Document, Volume 2: Description of Chemical Transport and Transformation Algorithms*. Office of Air Quality Planning and Standards, Research Triangle Park, NC. EPA 453/R-02-011b, available at http://www.epa.gov/ttn/fera/.
The current version of the TRIM.FaTE User's Guide is available at http://www.epa.gov/ttn/fera/trim_fate.html.

22. U.S. Environmental Protection Agency. 2001. *Risk Assessment Guidance for Superfund (RAGS): Volume I - Human Health Evaluation Manual (Part E, Supplemental Guidance for Dermal Risk Assessment), Interim.* Office of Emergency and Remedial Response, Washington, D.C. 20460. EPA/540/R/99/005, OSWER 9285.7-02EP, September 2001, available at http://www.epa.gov/oerrpage/superfund/programs/risk/ragse/index.htm.

23. U.S. Environmental Protection Agency. 1994. *Draft Technical Background Document for Soil Screening Guidance.* Office of Solid Waste and Emergency Response, Washington, D.C., December 1994. EPA/540/R-94/106.

24. U.S. Environmental Protection Agency. 1989. *Risk Assessment Guidance for Superfund: Volume I. Human Health Evaluation Manual (Part A).* Office of Emergency and Remedial Response. Washington, D.C., EPA/541/1-89/002, available at http://www.epa.gov/superfund/programs/risk/ragsa/index.htm

25. U.S. Environmental Protection Agency. 1986. *Guidelines for Mutagenicity Risk Assessment.* Risk Assessment Forum, Washington, DC. EPA/630/R-98/003; published in the Federal Register 51:(1185): 34006-34012, Sept 24, 1986, available at http://cfpub.epa.gov/ncea/raf/pdfs/mutagen2.pdf.

26. U.S. Environmental Protection Agency. 1989. *Exposure Factors Handbook.* Office of Research and Development, National Center for Environmental Assessment, Washington, D.C., May 1989. EPA/600/8-89/043, available at http://cfpub.epa.gov/ncea/cfm/recordisplay.cfm?deid=12464.

27. United States Environmental Protection Agency. 2003. *Guidance on Selecting the Appropriate Age Groups for Assessing Childhood Exposures to Environmental Contaminants (External Review Draft).* Risk Assessment Forum, Washington, D.C. September, 2003. EPA/630/P-03/003A, available at www.epa.gov/ncea.

28. U.S. Environmental Protection Agency. 2002. *Child-specific Exposure Factors Handbook (Interim Report).* Office of Research and Development, National Center for Environmental Assessment, Washington Office, Washington, DC, September 1, 2002. EPA-600-P-00-002B, available at http://cfpub.epa.gov/ncea/cfm/recordisplay.cfm?deid=55145.

29. U.S. Environmental Protection Agency. 1992. *Guidelines for Exposure Assessment.* Risk Assessment Forum, Washington, DC, May 1992. EPA 600Z-92/001, available at http://cfpub.epa.gov/ncea/cfm/recordisplay.cfm?deid=15263.

30. U.S. Environmental Protection Agency. 1998. *Guidelines for Ecological Risk Assessment.* Risk Assessment Forum Washington, April 1998. EPA/630/R095/002F. Available at: (http://cfpub.epa.gov/ncea/cfm/recordisplay.cfm?deid=12460)

31. U.S. Environmental Protection Agency. 1993. *Wildlife Exposure Factors Handbook.* Office of Research and Development, Washington, D.C. EPA/600/R-93/187. Available at: http://cfpub.epa.gov/ncea/cfm/wefh.cfm?ActType=default.

32. U.S. Environmental Protection Agency. 1992. *Framework for Ecological Risk Assessment.* Risk Assessment Forum, Washington, D.C., February 1992. EPA/630/R-92/001.

33. The current version of the TRIM.Risk$_{Eco}$ User's Guide is available at http://www.epa.gov/ttn/fera/trim_risk.html.

34. U.S. Environmental Protection Agency. 2003. *Generic Ecological Assessment Endpoints (GEAE) for Ecological Risk Assessment.* Risk Assessment Forum, Washington, D.C., October 2003. Available at: http://cfpub.epa.gov/ncea/raf/recordisplay.cfm?deid=55131.

www.ingramcontent.com/pod-product-compliance
Lightning Source LLC
Chambersburg PA
CBHW080640180526
45168CB00008B/3236